New technologies:
Their impact on employment
and the working environment

New technologies: Their impact on employment and the working environment

International Labour Office Geneva

ISBN 92-2-102975-1

First published 1982

The articles included in this volume originally appeared in the *Social and Labour Bulletin* (published by the ILO).

ILO publications can be obtained through major booksellers or ILO local offices in many countries, or direct from ILO Publications, International Labour Office, CH-1211 Geneva 22, Switzerland. A catalogue or list of new publications will be sent free of charge from the above address.

Printed in Switzerland

PREFACE

"Chips to change offices ...", "Office automation takes over the factory ...", "Microelectronics revolution arrives ...", "The robots are here and ready to help ...". These and similar headlines fill the papers and provide a fertile ground for speculation. To what extent will new technologies alter current working patterns, affect employment and even change society as we have known it up to now?

Because so many questions still remain unanswered, there has been extensive readership response to the technology chapter in the ILO's quarterly publication the Social and Labour Bulletin (S.L.B.). To meet this response, it was decided to publish a separate reprint of all Bulletin articles dealing with the repercussions of new technology since 1979. These articles do not represent an exhaustive or definitive picture of the "state of the art"; they do, however, show the context in which the current debate on the implications of new technology is taking place. They illustrate how both the people and organisations affected are reacting and describe national initiatives, international preoccupations, industry reactions, safety and health implications and the views of users and academics.

Listed alphabetically by country, the articles are grouped under six main headings: general policy issues; privacy protection; trade union viewpoint; labour relations and collective agreements; employment, training and qualifications; work environment.

Reporting is based on the news media and original source documentation supplied by the various interested organisations.

Although relatively few documents have been received from employers' organisations, the views of the employers can be found under the various headings. As the headings are, of necessity, fairly general, many of the articles touch on other aspects than those under which they are specifically listed.

This publication is just one dimension of the ILO's over-all programme on new technologies:

- "Fragmented jobs in offices" and "Automation and work design" are the titles of two new ILO studies dealing with conditions of work and now nearing completion. The first looks at data entry jobs, including word processing, where microprocessors have completely altered traditional information handling. The second examines impacts in industry, covering numerically controlled machines, industrial robots and computer process control. Both include country and case studies, comparative analyses and suggested strategies, policies and practices for improved work organisation and job content. Future plans for 1982-83 include the preparation of guidelines for introducing new technologies at enterprise level and a Meeting of Experts on automation, work organisation, work intensity and occupational stress.

- The Employment and Development Department has published the first study of a series on the employment implications of microelectronics (J. Rada, The impact of microelectronics, ILO, Geneva, 1980). Further studies in this series are under way, dealing with the effects of new technologies on the textile and garment industry, on the electronic industry and the international division of labour, and on women's employment. These studies will be published in the course of 1982. Further publications dealing with other industrial sectors are planned for 1982-83.

- Information technology is also a regular component in the programmes of the ILO Training Department.

- Several recent reports issued by the ILO Industrial Committees have also touched on various aspects of this question. The report prepared for the Second Tripartite Technical Meeting for the Printing and Allied Trades deals exclusively with the repercussions of changing technology on skill levels (Training and retraining needs in the printing and allied trades, Report II, Geneva, 1981). Another report prepared for the Advisory Committee on Salaried Employees and Professional Workers, also deals with employment and working conditions (The effects of technologies and structural changes on the employment and working conditions of non-manual workers, Report II, Geneva, 1980).

- The International Labour Review, a leading ILO periodical appearing every two months has devoted several articles to this subject.

- The Social and Labour Bulletin will continue reporting on this important question.

TABLE OF CONTENTS

TABLE OF CONTENTS

TABLE OF CONTENTS

IMPACT OF NEW TECHNOLOGIES

WORKING ENVIRONMENT

GENERAL POLICY ISSUES

AUSTRALIA

Committee on technological change established

An interdepartmental committee to study the impact and implications of new technology was set up by the Government in the last quarter of 1978. It will study not only the problems created by new technology but also the consequences of not introducing it.

Source: Victorian Employers' Federation Report, Vol. 8, 15 Sep. 1978, p. 3.

Special unit monitors changing technology scene

In a speech given on 17 October 1979, the Minister for Industrial Relations, Technology and Energy of New South Wales announced that the recently formed Technological Information and Research Unit had become fully operational. (At the national level, a tripartite Committee of Inquiry into Technological Change was set up in 1978, see S.L.B. 1/79, p. 1).

Indications were that within the next two years more than 30,000 people in the State might be affected by technological change. The new unit will therefore (1) obtain information about the interaction between technology and society, and especially about the effects of technological change on the level and quality of employment; (2) analyse this information and the issues relating to technological change; (3) have a disseminating role, not only to the Government but to the whole community so as to promote informed and constructive public discussion; (4) suggest priorities for government action.

Although concentrating on levels and quality of employment, the unit will also examine the industrial relations implications of advanced technology and help management and unions to resolve problems with a minimum of hardship for the workforce.

Another task will be to develop soundly-based manpower planning strategies and recommend changes in training and retraining practices. It will, for instance, provide information on technological change that will allow the Department of Technical and Further Education to update courses and develop new ones where necessary. Vocational counsellors and school guidance officers will get information on the latest developments. It will also consider the development of job creation schemes and alternative occupations for persons displaced by such change.

Source: Speech by the NSW Minister for Industrial Relattions, Technology and Energy, on 17 Oct. 1979 (reprint, 6 pp.).

The Myers report on technological change

The long awaited report by the Committee of Inquiry into Technological Change (see S.L.B. 1/79, p. 1) was submitted to the Government in April 1980 and subsequently released to the public.

The report was commissioned by the Federal Government "to examine, report and make recommendations on the process of technological change in Australian industry in order to maximise economic, social and other benefits and minimise any possible adverse consequences". It was prepared by a 3-man committee headed by Professor Rupert Myers who gave his name to the so-called "Myers Report". The other two members represent employers and workers respectively.

In carrying out its brief, the Committee was asked to identify technological change and its impact, to examine overseas experience and to review the effectiveness of government policies and programmes in facilitating the introduction of new technology.

The 1,555-page Myers Report comprises four volumes. The first reviews the issues involved in technological change and sets out the main recommendations. Volume II contains more detailed descriptions of likely technologies and their application in Australia and Volume III, a review of government policies and programmes. The final volume contains selected papers on key industries and issues.

The report urges industry to keep up with world technological development saying "that it cannot find 'evidence' that the current high level of unemployment is attributable to technological change". It sees general economic conditions playing the main role, and the failure of an industrial country to sustain technological progress as likely to cause serious structural unemployment and loss of competitiveness.

As well as defining the means by which venture capital could be made available for small innovative enterprises, the Committee recommends that the Australian Industrial Research and Development Incentives Board (AIR/DIB) be restructured to become autonomous and to have its charter expanded to include funding of public interest projects. The AIR/DIB is a government board which grants money for research and development projects. The Committee thinks that such funding would create more jobs than simply increasing government money for research.

Suggesting as a possible model the West German venture capital bank, Deutsche-Wagnesfinanzierungs-Gesellschaft (WFG), which is a limited company funded by banks, other credit institutions and the Government, the

Committee recommends government sponsorship of a private venture-capital corporation to provide risk capital to individuals and to small and medium-sized enterprises to promote the production and marketing of promising inventions and innovations. The government contribution should include a guarantee of loan funds up to a maximum of AUS$25 million over 15 years which would be matched by private investors (1 US dollar = 0.86 Aus. dollars). The Government should also have wider powers to fund research that is the public interest.

While admitting that technology can reduce activity in specific sectors, on the whole the Committee sees a strong link between technological change and economic growth and resultant rising incomes which lead to the creation of new employment opportunities elsewhere in the economy. A major recommendation is that unemployment benefits be supplemented by a "social safety net" designed to assist people to adapt to change and to help those displaced by the new technology. Such a safety net would provide for compensation payments, after retrenchment, of 60% to 75% of salary for between one and six months, depending on length of service. Extra allowances are suggested for those with dependants.

Moreover, to assist individuals who are to be retrenched through no fault of their own to find alternative employment, the Committee believes that appropriate provisions should be made in collective agreements covering periods of notice, notification to the employment service, time off for job interviews, monetary compensation and assistance to find alternative employment.

In addition, much more should be done by the Government in the field of training to: (1) identify likely technological changes; (2) assess the effects of such technologies on skill requirements; and (3) provide the relevant basic and further training.

Among other recommendations both to promote and enhance the acceptance of technology were the following:

- to set up a technology awareness programme to promote technological change and explain its effects on the economy, industry, the labour force and society;
- to enhance the capacity of the union movement to study and respond to technological change. The Trade Union Training Authority (TUTA) should draw up recommendations and additional government funding should be agreed where significant needs are identified. The Australian Council of Trade Unions (ACTU) should be invited to set up a special unit to specialise in technological change and the Government should provide a grant towards its establishment of operating costs;
- that legislation relating to the organisation of unions be amended to facilitate the formation of larger industry-based unions;

- the Australian Science and Technology Council (ASTEC) should set up a standing committee with representatives from both sides of industry, the scientific world and Government to monitor and report on the effects of technological change at national level;
- to ensure a Commonwealth/State consultation at the ministerial level through a special Council on Technological Change;
- to introduce appropriate legislation to ensure the protection of personal privacy in relation to new technology;
- to provide for wage margins that more properly reflect skills and responsibilities;
- to upgrade working conditions via a "bureau of the working environment". Such a bureau should aim at making workplaces more pleasant and more safe, as well as at enhancing the quality of work through attention to factors such as work organisation and patterns of work time. The proposed bureau could consolidate existing guidelines on safety practices and issue supporting codes from time to time. The Committee considers that the general question of the physical working environment is very important in relation to technological change;
- to consider how the tax system might be influencing the rate of technological change and to devise means of removing disincentives to industry to innovate (e.g. a review of current depreciation rates, eventual abolition of the payroll tax).

In presenting its report the Committee recognised that there are inherent limitations in the extent to which predictions can be made. In many cases the range and complexity of technologies, and their rapid evolution, make predictions hazardous and uncertain. In others, the rate of introduction depends on economic and social factors that themselves are unpredictable.

In a detailed government response to the report in Parliament on 18 September 1980, the Minister for Industry and Commerce strongly endorsed the central conclusion of the Committee that it is essential for Australian industry to keep up with technological developments. It also accepted the general recommendations for fostering a more consultative approach and on training, research, public interest projects, wage margins more properly reflecting skills and responsibilities, a technology awareness and monitoring programme, Commonwealth-State consultation, and privacy protection.

However, the proposals concerning redundancy arrangements and temporary income maintenance raise complex and major issues for the Government, employers and unions alike. Before taking a final decision the Government will hold talks with both sides of industry. Government sponsorship of a private venture capital corporation has been referred to a Committee of Inquiry into the Australian Financial System, while issues relating to the tax system will require consideration by state governments.

The Australian Council of Trade Unions (ACTU) says the report recommendations are an improvement over current practice but are in many cases over-simplistic. While the recommendations will need a close and detailed examination by the trade union movement, the ACTU notes the positive approach to minimum standards of employment security, including income maintenance, for employees affecting by technological change and to the principle of greater tripartite consultation. The recommendations demand that the Government re-examine its currently inflexible approach to public spending and the ACTU sees future trade union co-operation with the introduction of new technology as heavily influenced by the government response to the recommendations.

For its part, the Confederation of Australian Industry supports the conclusion of the Myers Committee that the implementation of new technology must be encouraged to maintain the competitiveness of industry. However, it sees the concentration on changes in the industrial relations system as a solution to the problems involved, to be naive in the extreme. While at the same time, solutions to the serious training problems identified in the report have been dodged.

Source: Australia: Technological Change in Australia: Report of the Committee of Inquiry into Technological Change in Australia, Vol. 1: Technological Change and its Consequences, including discussion of processes of change, conclusions and recommendations for maximising benefits and minimising adverse consequences (Canberra), 1980, 257 pp.
Government response to the report of the Committee of Inquiry into Technological Change in Australia, 18 Sept. 1980, 9 pp. (mimeo.), plus detailed response to each recommendation, 17 pp.
ACTU Executive Decision: CITCA Report, Sept. 1980, 4 pp. (mimeo.).
Confederation of Australian Industry: Media Statement, 31 July 1980.

CHINA

National computer development

The first national co-operative society of computer users was set up on 2 July 1980 in China to promote innovation and experience exchange in this field. In the same month, China's first computer service company, providing maintenance and programming services as well as training technical personnel, also started operating.

A total of 2,615 computers, including 318 of foreign make, are actually being used in scientific research, economic management and other sectors. The first computer forecast of the country's population trends in the next 100 years was carried out recently.

Source: Xinhua News Agency: Information quoted by Daily Report, People's Republic of China (Springfield, USA), 29 May, 17 June, 3 and 23 July, 1980.

CZECHOSLOVAKIA

Electronics is a priority sector

According to a 10-year target programme approved in December 1979, the electronics sector is expected to play a key role in Czechoslovakia's economy, with about 8 billion crowns being invested for its development by 1985 (1 US dollar = 11.94 Cz. crowns).

While, by 1990, the sector's planned output will expand by some 45%, its labour force will rise by only 2.7% (from 220,000 employed at present).

At the same time, electronic devices and specifically microprocessors are considered as a means for solving labour shortages and are expected to replace over 300,000 workers in other industrial sectors by 1990. For example, it is estimated that the use of electronics in the textile industry (development of open-end spinning frames, weaving looms, knitting, sewing and finishing machines with electronic components) would provide for a 50% to 70% labour saving.

Source: Czechoslovak News Agency (CTK): Czechoslovak Digest, Background Information (Prague), No. 7, 1981.
Rude Pravo (Prague), 23 Oct. 1980.

FRANCE

Computerisation: new agency publishes action programme

A five-year action programme on computerisation, to be introduced by stages, was approved by the French Government on 6 December 1978. An initiation and training programme is to be introduced in secondary schools, where 10,000 micro-computers are to be installed. Teacher training in this new discipline is to be expanded and systematic basic training in data processing methods will be introduced in higher education establishments for non-specialised students.

A budget of 2,250 million francs (1 US dollar = 4.15 French francs) will be allocated to the Ministry of Industry for the promotion of data processing in education, industry and small and medium-sized undertakings, for the development of new computer applications in industry, for office equipment and data banks so as to improve the competitive position of French undertakings. Special attention will be given to the computerisation of production processess in the automobile industry, machine tools, printing, banks and insurance companies. An agency will be set up in the Ministry of Industry to co-ordinate state action and allocate funds.

The Ministries of Labour and Industry and other government departments will carry out studies to determine the influence of computerisation on working conditions.

This programme was adopted a few months after the publication of an in-depth study on the computerisation of French society, commissioned by the Government (see S.L.B. 3/78, p. 207).

Source: Le Monde (Paris), 1, 2 and 8 Dec. 1978.

Teletext versus paper: an all-out contest?

This was the subject chosen by the French section of the International Institute of Communications (IIC) and the Association Téléqual for the meeting which they organised at Paris on 12 November 1979 under the sponsorship and with the participation of the Minister of Industry, the Minister of Culture and Communication and the Secretary of State for Postal Services and Telecommunications. Attended by some 400 participants and 40 top-level government and industry executives, the meeting gave technicians, legal experts, press and publishing representatives an opportunity to exchange views on the topical issue of the future of the written press ("paper") faced with the increasing application of telematics (telephone + television + computer).

The most salient points of a 15-page working paper are reviewed below.

In less than a year the General Directorate of Telecommunications (DGT) will deliver free of charge to 2,000 households in the Ille-et-Vilaine area a small 20 cm television screen and an alphanumeric keyboard, both connected to the telephone. By operation of the code which each subscriber will receive, the screen will automatically show any telephone number he wishes to ascertain. A unit charge will be made for this service. Because of the savings in paper which this electronic "telephone directory" system will allow and the relatively low cost of the mass-produced apparatus, the widespread introduction of automatic screen reading will cost less than the use of paper telephone directories.

At the same time, another "Télétel" experiment is being set up by the postal and telecommunication administration (PTT): by the end of next year 3,000 households in Velizy will be equipped with telecommunication terminals enabling them after a simple telephone call to receive on the screen of their television set a wide range of information (chemists on duty, reservation of theatre seats, airline and railway tickets, weather reports, various documentary data, etc.). The technical possibilities of this screen-telephone connection are limitless (including classified advertisements and even daily newspapers if the network is connected up with a press agency, etc.).

It is here that the press is threatened. Telematics will force readjustments in four different areas:

(a) In the internal organisation of each press undertaking. Readjustments will be needed within the editing-printing system. Since the typewriter will give way to a screen photo-typesetter whereby typing, correcting and (soon) page-setting can be made directly, the very function of the editor is at stake, as is the function of the printing staff. Telefacsimile is also changing the relations between the editorial head office and correspondents in the provinces.

(b) In the sending and distribution of messages. By electronic mailing (telephone + screen and printer) it is possible to send by telecopier not only individual letters but documents for large-scale circulation having all the characteristics of present-day newspapers. Also, with the use of the "inter-active video text" whereby a dialogue can be established between the reader and the machine, each user may obtain a "personalised" newspaper, i.e. one geared to his specific needs and tastes. A new type of mass media therefore comes into being.

(c) In relations with the telephone. With the possibility of transmitting images by telephone and of adding software and a printer to the screen, the PTT monopoly has come to represent a direct threat to the written press. It is therefore essential to establish clearly the respective roles that should be played by government and the private sector, if possible for the benefit of the individual.

(d) Memories. The average user, by means of a keyboard, a viewing console and a telephone line, will soon have direct access to memories whose capacities are constantly increasing while the cost of memorisation is constantly decreasing (in five years' time it is expected to fall from half a centime per information bit, i.e. per impulsion, to one-thousandth of a franc; 1 US dollar = 4.10 French francs). Here it is not only the press but books that are threatened.

The speakers at the meeting (who included Mr. S. Nora, co-author of the Nora-Minc report on the computerisation of society, analysed in S.L.B. 3/78, p. 207) reviewed the various political and technical aspects of the applications of telematics. (The general conclusion was that the various systems of communication and transmission of the written word now available should tend to complement rather than compete with one another.

They also considered the legal aspects of protecting signs and symbols, monitoring message content, network ownership, copyrights, etc. in connection with such applications.

A subject dealt with at length was the future of broadcasting monopolies (in France, the PTT and the television broadcasting system) the dual nature of which will soon cease to be warranted and which should give place to a public service.

Another meeting was convened at Paris on 9 and 10 December 1979 by some 12 left-wing magazines on "computerisation versus society". The participants in this meeting considered that the various computerisation projects in hand at present aim at transforming social relations, and they stressed the dangers involved: cultural standardisation, crushing of the individual and growing control of society which the shaky safeguards provided for by the 1978 Act on computerisation and individual freedoms will not be sufficient to check.

Since the trade unions, social movements and political parties are not prepared to cope with such developments, it was recommended that a body to be known as the Centre for Information and Initiatives on Computerisation (CIII) should be set up to establish direct links between workers in the various sectors concerned with computerisation and in particular between computer experts and users.

Source: International Institute of Communications - Association Téléqual: "Le match du siècle: télétexte contre papier". Working paper and proceedings of the meeting (Paris), Nov. 1979, 17 pp. and 103 pp.
CIII: Supplement to "Partis pris", No. 16, Commission paritaire (Paris), 2 pp.
Le Monde (Paris), 14 Dec. 1979.

FEDERAL REPUBLIC OF GERMANY

Industrial society and technological challenge

This was the theme of the 5th Scientific Forum of the Institute of the German Economy (DIW) which took place in Cologne (FRG) in January 1981. The purpose was both to discuss the nature of new technologies as well as their effects on society and the economy and to give the floor to those who question the direction which recent technological developments are taking.

The forum was to highlight the dilemma of industrial societies. On the one hand, fear and apprehension regarding uncontrolled growth which recklessly consumes vital energy sources, pollutes air and water and destroys the environment; on the other hand, economic progress, technology, research and development which are the very basis of existence. Without sufficient growth and without adequate sources, our closely-knit economic and social system would soon collapse. Without innovation and rationalisation which increase not only productivity but the quality of working life, social tensions would soon reach a point of no return.

The FRG is poor in raw materials. If the new generation's future is to be safeguarded, it is vital to keep pace with technological progress and to strive for heightened creativity and innovation. The forum emphasised that it would be irresponsible to focus exclusively on the negative aspects of industrial society. This would indeed undermine the drive and optimism of the young generation. On the contrary, the direction to take is to use technical progress to ensure jobs and living standards, to improve the employment structure, to protect the environment, to safeguard social institutions and to be of effective help to developing countries.

At the same time, it would be illusory to pretend that technology advances are without risks, without negative consequences. Both its positive and negative aspects are and have been part of men's destiny. A critical attitude must be based on facts and experience and not expressed merely for the sake of opposing technology. The manifold consequences of technology have to be recognised and mastered. Resignation to an all powerful "technostructure" would be fatal. The most visible and vocal expression of growing uneasiness regarding the advance of bureaucratic and technological structures are the ecological and alternative movements. They cut across traditional political and social groupings which are considered incapable of solving the problems facing the society. An illustration is the vehement public debate over nuclear energy. The alternative proposals are not hostile to new technologies per se - they aim at the system behind them. One issue is participation in decisions concerning the use of technology. There is a variety of alternative scenarios. The forum's concern was to point out those that seem politically relevant and socially tolerable.

Source: DIW: Industriegesellschaft und Technologische Herausforderung, by W. Schlaffke and O. Vogel, Deutscher Instituts-Verlag (Cologne), 1981, 400 pp. (DIV-Sachbuchreihe, Vol. 25).

Rationalisation and automation in the 1980s

Rationalisation is one of the central themes today both in social matters and in business and industry. The focus is particularly on microelectronics - is it a "job-killer" or a "job-crumpler"? An article in the monthly review of the Federal Labour Ministry attempts to answer these questions from information collected during a meeting at the Evangelical Academy in Baden late in 1980.

The author predicts that in the next ten years practically feasible new technologies will mainly be in the following areas: automatic production control and stock control, data techniques and text processing in offices, micro-electronic techniques, new media techniques (e.g. satellite communications), new energy sources and uses, laser technology, and environment-monitoring techniques.

As examples of the new techniques mentioned are a micro-processor that replaces 350 mechanical parts in a sewing machine and the fact that the assembly of an ordinary mechanical watch requires about 1,000 different operations, whereas an electronic watch is assembled from five basic components: the case, a quartz crystal, a diode, the integrated circuit and the battery.

It is estimated that offices and institutions will feel the effects of the new technologies by 1985, press and information will be significantly affected by 1990, automation in industry which is already taking place will continue until 1985 - robots on the other hand are still in their infancy and will initially gain in importance by the end of the 1980s.

There appears to be no reason for optimism as far as the employment effect of the new technologies is concerned. The prediction is that the 1980s will witness an increase in productivity coupled with a drop in the number of persons gainfully employed. According to the Federal Government's annual industrial statistics of 28 January 1981, this year will see a fall in real industrial growth from 0 to -1% and an average unemployment figure of 1.19 million.

Some of the effects have already been observable in different enterprises. In the Volvo works, a manual spot-welding line was replaced by an automatic one and as a result the number of welders was reduced from 100 to 20. A newspaper print shop introduced a computer-controlled setting machine, reducing the manning per shift from 59 to 29. The new technique also increased the number of characters set per hour from 25,000 to 8 million.

Micro-processors also affect the level of employee qualifications. The tendency is towards deskilling of jobs which require adaptation and in some cases retraining of the part of the workforce affected.

As far as industrial growth and job security are concerned, the author is careful in both predictions and recommendations. He does, however, point out that over the last 80 years technological development has time and again led to a reduction of working hours in the country from about 60 hours per week in 1900 to the current average of a 40-hour week.

In the author's opinion, the solution to the problems that may result from rationalisation and automation in the 1980s can only be found through a broad policy shaped by the State, the negotiating partners and industry if the country is to keep in the running as far as international technological development is concerned and at the same time overcome the negative effects on employment.

Source: FRG: Federal Ministry of Labour and Social Affairs: Bundesarbeitsblatt (Bonn), No. 4, Apr. 1981, pp. 5-11.

Industrial robots re-examined

Industrial robots are increasingly used in production says the Association of German Engineers (VDI) and although the general opinion is that industrial robots take away jobs, this is not quite correct. According to the VDI, industrial robots (IR) are a very useful and flexible aid in production and can actually make jobs more humane by removing the need for workers to engage in tiring and boring work.

IRs can feed, take off, stack, sort, tidy, assemble, screw, grind, spray, check, drill, spot and line weld, etc. They do it not only with the required accuracy and fast workpace but the fact that they can do it in hot, humid or dusty rooms brings many advantages.

The great advantage of IRs is that they can also carry out dangerous operations. As safety at work is a high priority, the manufacturers, users, joint associations for accident prevention and trade unions agreed to take part in a technical meeting on the problems with a view to creating humane working conditions through the use of industrial robots.

The meeting was scheduled to deal with the introduction of industrial robots and allied appliances, their construction, installation, rules governing their use, their social effects, issues related to ergonomics, work organisation, wages, and the necessary adjustment to legislation. The reasons for the introduction of IRs from the employers' point of view was also on the agenda.

Source: FRG: Federal Ministry of Labour and Social Affairs: Bundesarbeitsblatt (Bonn), No. 4, Apr. 1981, p. 34.

JAPAN

Special committee reviews impact of microelectronics

Technology-intensive countries are increasingly concerned about adverse effects on employment. For this reason the Ministry of International Trade and Industry sponsored the setting up of a Special Committee on the Impact of Microelectronics on Employment. The Committee recently commissioned a fact-finding study on microcomputers and their employment impact.

Carried out by the Research Institute of the Japan Management Association, the study attempted to analyse the general trend for popular diffusion of microcomputers. Interviews were then carried out with corporate executives from 20 microcomputer-producing companies to try and identify future developments affecting industrial products, office machines and computer products. Some tentative estimations made as a result of the study are outlined below.

The microcomputer market falls into two main categories: the suppliers engaged in designing, making and selling the products and the users engaged in applying (or potentially applying) microcircuits to products and processes of all kinds.

The impact of microcomputers and associated products on employees of the suppliers are summarised as follows:

(1) There is an urgent need for software development personnel as a result of increasing demand for sophisticated and custom-made software.
(2) In manufacturing the size of the workforce has remained stable due to an increase in demand. However, in the long term a decrease is a foregone conclusion due to facility integration. Moreover, owing to process innovations, productivity has increased and demand for skilled labour decreased. In 1985, the workforce will reach an estimated total of 2.43 million through natural expansion. However, microcomputer application is expected to reduce nominal employment by a maximum of 480,000, or a minimum of 210,000.
(3) In the marketing area, personnel has increased slightly.
(4) In the maintenance area, no change has been observed.

As for the impact on employees of the users caused by microcomputers and associated products, it is generally forecast that substantial numbers of the existing labour force will lose their jobs because of the labour-saving aspects of the equipment and that a substantial shortage will occur in systems and software personnel.

Advantages offered to equipment users include: automation of analysing, measuring and inspection by microcomputer-intelligent equipment, high precision, reliability, flexibility and simplification of work. This means that skilled workers are no longer needed. Part-time or unskilled labour can operate the equipment and so lower operating costs.

The following table shows the impact of microcomputers on the workforce of suppliers and users:

Suppliers					Users		
Development, design	Mfg.	Sales, maintenance	Spare parts	Product area	1	2	3
+	*	+	-	Analytic, measuring	NA	-	NA
*	-	*	-	Process control	-	-	NA
+	-	*	*	Manufacturing automation	NA	-	-
+	-	+	NA	Office Machines	NA	x	x
+	-	+	-	Commercial machines	+	-	x
+	-	*	-	Watches	NA	NA	NA
-	-	*	NA	Calculators	NA	NA	NA
+	x	x	x	Sewing machines	NA	NA	NA

\+ Increase
\- Decrease
* No change
x Unknown
NA Not applicable

1 Engineers
2 Skilled workers
3 Unskilled workers

Source: Japan: A fact-finding study on the impacts of microcomputers on employment, Summary prepared by a working group on behalf of the Special Committee on the Impacts of Microelectronics on Employment (Tokyo), 10 Nov. 1979, 37 pp.

Automation in Japanese industry as seen by French industrialists

A study group from the French Technical Centre for the Mechanical Engineering Industries (CETIM) visited Japan in 1979 to investigate the automation of manufacturing methods in mechanical engineering. Its report was considered at a one-day meeting held in March 1980 to study this question. The major observations and conclusions of the meeting are given below.

Japanese large-scale manufacturing industries work virtually without stocks, using subcontractors to supply their assembly lines with all kinds of components, and relying on automatic batch-processing methods which can be adapted at very short notice.

The industries themselves have developed flexible assembly lines which can batch process products with fairly wide degrees of variation. Production is modified by electronic control without stopping the assembly line. One of the factories visited, for example, was able to manufacture thermometers in 50 different versions. The flexible assembly line was programmed to obey nearly 200 commands to change the model in one day without stopping either the assembly line or production. A second's down time was enough to carry out the required change at each work post.

These short production runs based on flexible automation are made possible by the numerical control machine. In 1978 there were already more than 7,300 machines of this kind in Japan, compared with 500 in France. The machines are fed and tended by robots: manual manipulators, devices with fixed or variable sequences, robots with memories, etc. In 1977 over 30,000 robots were operational in Japan, 35% in the automobile industry, 25% in the electrical industry and 20% in mechanical engineering. Over half of these robots operated on fixed sequences and 27% on variable sequences. They are generally capable of three movements, usually two linear movements and one circular movement. A total of 130 manufacturers produce machinery of this kind, 40 of whom belong to the Japan Industrial Robot Association (JIRA), a very active body which is subsidised by the Ministry of International Trade and Industry (MITI).

Research in this field enjoys substantial government support mainly because: the young population of Japan is tending to decrease, while the level of education is rising; wages are rising to levels comparable with the highest levels prevailing in the West; safety regulations are more stringent than before; and the demand for high quality is increasing.

The MITI is responsible for a major national project, the flexible manufacturing system complex, i.e. an entirely automated "unmanned workshop". Studies and work on this project, which is already well-known throughout the world, are at an advanced stage, but the assembly of the installation as a whole has not yet begun. Its cost is estimated at 12,000 million yen (1 US dollar = 225 yen). The workshop, which comprises a number of subsystems, each of which corresponds to a specific function (shaping, assembly, laser treatment, automatic control and fault detection) is designed for the manufacture of engineering components and units weighing less than 500 kg in batches of not more than 300. Automation is expected to lower manufacturing times by at least 50% compared with conventional workshops.

Japan at present manufactures 40% of the robots sold throughout the world, which places it well in the forefront of world manufacturers. Most of these robots are used in Japan, but in 1985 exports are expected to account for 20 to 30% of production. For the time being, stress is laid on simple, inexpensive robots for the loading and unloading of machine tools, punch and stamping presses and plastic injection-moulding presses. Applications in other sectors will no doubt be developed in the near future (nuclear energy, exploitation of the sea bed, assistance to invalids, etc.).

It may be recalled that the MITI had also sponsored a study on the impact of microelectronics on employment which was analysed in S.L.B. 2/80, p. 139.

Source: CETIM: "Le Japon de 1979 et l'automatisation des fabrications mécaniques". Working papers submitted to the study group at Senlis, 27 Mar. 1980, CETIM Informations (Paris), No. 62, 13 pp.

USSR

Enter the robots

In August 1980 the Central Committee of the Communist Party of the Soviet Union published a resolution urging responsible ministries and agencies to accelerate the development and production use of automatic manipulators (industrial robots) which are expected to help in solving the manpower supply problems of the 1980s. (There are about 60 different types of automatic manipulators actually produced in the USSR; each device, on average, saves the labour of 3 workers.)

The resolution provides for a comprehensive national research and development programme for 1981-1990 aimed at introducing more robots in engineering, mining, iron and steel, non-ferrous metallurgy, construction, transportation, agriculture and the food industry. It includes measures to improve the supply of qualified personnel for robot manufacturing and servicing. The savings derived from use of robots will remain at the disposal of undertakings concerned, to provide additional incentives for their staff.

Particular attention should be given to introducing robots in workshops with unhealthy and harmful working conditions, especially in hot-dip plating, varnishing and similar operations where, in the next five years, human labour is expected to be completely replaced by automatic manipulators.

Source: Pravda (Moscow), 9 Aug. and 19 Sept. 1980.
Mekhanizatsia i avtomatizatsia proizvodstva (Moscow), No. 2, 1980, pp. 5-13, 31-34.

UNITED KINGDOM

Microprocessors: unions reticent as Government boosts development

A code of practice for the introduction of word processing machines was suggested in a report issued in December 1978 by the Association of Professional, Executive, Clerical and Computer Staff (APEX). The union is worried about the effect of automatic typing systems which have a memory. The systems improve productivity by eliminating the need for redrafting and allowing for rapid correction of errors. One of the latest systems can edit, arrange text in any order, sort names on lists, locate names or words and carry out limited mathematical operations.

The APEX report highlights certain dangers seen by the union ranging from possible health hazards and unemployment to the creation of monotonous jobs. It points out that unemployment among clerical workers is above the national average, and that the widespread introduction of word processors will aggravate this situation.

APEX would like to see more research on the possible health hazards (eye strain, possible radiation from defective screens, the effects of isolation in a word centre without outside contacts), although recognising at the same time the positive effects of such systems: release from tedious and repetitive work to more responsible and interesting tasks.

Union concern is voiced at a time when the Government is investing large sums of money in supporting microelectronics. Initial funding of 15 million pounds (1 US dollar = 0.51 pounds sterling) in July 1978, to promote the widespread use of microprocessors, has since been increased to 55 million. The new microprocessor scheme complements government support for the microelectronics industry which will cost 400 million pounds over a 3-year period. This, in turn, is expected to generate further investment of some 250 million pounds. Moreover, says the Government, job reductions should be matched by the creation of new jobs.

The government programme involves: making people aware of the potential of microelectronics; training and retraining workers; reorientation education; direct financial support to industry; and gearing public sector purchasing to these new developments wherever possible. A new unit is being set up in the Department of Industry to promote microelectronic technology while the Department of Employment has a study group looking at potential manpower implications.

Source: Financial Times (London), 27 July, 23 Oct., 1 and 7 Dec. 1978.
Trade and Industry, HMSO (London), 7 July 1978.
Department of Employment Gazette, HMSO (London), Oct. 1978.

Industrial robotics: research impetus

A 5-year research and development programme in industrial robotics costing some 500,000 pounds a year (1 US dollar = 0.4 pounds sterling) was launched by the Science Research Council on 24 July 1980.

Carried out by university/industry partnerships, the research aims to "leapfrog" the present generation of robotic devices. The first partnerships under the programme will be dealing with the assembly of electronic circuit boards and small electronic motors; free-roving driverless warehousing trucks; and robot computer software simulation.

Source: British Information Services: Survey of Current Affairs (London), Vol. 10, No. 8, Aug. 1980, p. 252.

Equal opportunities: Impact of information technology

The rapid introduction of new electronic technologies - especially word processors - means that the demand for certain types of skill will diminish, while new opportunities will open up. However, the jobs that will disappear are those traditionally held by women (e.g. secretarial and clerical jobs), while the new jobs will be at a more senior level, in fields where men are generally dominant. This is the main conclusion of a recent report issued by the Equal Opportunities Commission (EOC).

Estimating that about one million women work as secretaries and typists, the report predicts job losses of 21,000 by 1985 and of 170,000 by 1990. The report estimates that for each word processing unit installed, one third of a typist's job is lost. The picture is even more bleak for clerical work. It is forecast that up to 40% of clerical jobs may be lost during the 1980s. The report cites examples of jobs lost - in one instance amounting to as many as 800 jobs.

However, the staff displaced by the new technology will not be absorbed by the new jobs it will also create. First, the report suggests that the jobs created will be less than the jobs lost. Second, the new jobs are in fields where male employment is predominant (e.g. sales, maintenance, computer programmers, engineers). The report estimates that in 1979 6,000 typing jobs were lost, while around 800 extra staff were recruited by word processing manufacturers. Around 60% of these new jobs were for sales representatives, and are likely to have been filled by men, while the rest were probably filled by women. The main recommendation of the report calls for more training for women so that they will be suitably qualified for the vacancies that exist.

As far as job satisfaction was concerned, the report comes up with contradictory findings. Some typists enjoyed learning a new skill, and the time spent in actual typing decreased. Typists' salaries rose, but not to the level paid to secretaries. An obstacle to career advancement was that word processing operators move out of the secretarial hierarchy and are less likely to be promoted to top private secretary posts.

Although, in theory, automation increases options for working part-time or from home, the report found no moves in this direction.

Source: United Kingdom: EOC: Information technology in the office: the impact on women's jobs, by Emma Bird (Manchester), 1980, 100 pp.
Industrial Relations Review and Report (London), No. 235, Nov. 1980, pp. 9-10.

Technology bridge with Third World

The recently formed UK Council for Computing Development aims to narrow the computing gap between Third World and Western countries. Encouragement for the proposal had come from the Government and industry in the UK.

Ghana and Argentina have already asked to consult the Council, and international agencies such as the World Health Organisation had welcomed it, particularly since there had been an increase in demand for help with informatics in health projects.

Source: Computer Weekly (London), 21 Mar. 1981.

UNITED STATES

Transborder data flow: jobs, privacy and national sovereignty

According to a recent opinion survey made for a congressional subcommittee in the USA, there is substantial support among policy makers and businessmen around the world for restricting transborder data flow (see S.L.B. 1/80, p. 13 concerning Council of Europe's proposed international convention).

Carried out by the Transnational Data Reporting Service, Inc. (TDRS), the survey tried to provide an overview of current thinking and policies on international data flow issues. There were 561 replies from 67 countries.

The responses, representing government, business, academics and other interests, showed that: (1) 61% of respondents concur with the statement that "the absence of national capabilities to meet a country's data-processing needs compromises its national sovereignty"; (2) 76% think data communications policies should be part of an over-all government plan; (3) 40% believe that "national legislation will significantly reduce the growth of transborder flows of information".

TDRS say the responses suggest that transborder data flow and related issues of national sovereignty, unemployment and personal privacy will increasingly occupy important places on national, social and economic policy agendas.

The survey comes at a time of increasing concern in Europe and elsewhere that the US computer industry can damage their respective economies. A Canadian government report, for example, has charged that Canada loses 35,000 jobs annually because US companies process Canadian data that otherwise would be processed by Canadian companies.

Source: Computerworld (Framingham, Mass.), 7 Apr. 1980, pp. 17-19.

EUROPE

Seminar on electronic data base networks

The Savant Institute in Munich (FRG) organised, from 8 to 12 January 1979, a seminar on "Distributed Processing and Data Base Networks in the 1980s", which was attended by some 180 senior managers from public administration and business in Austria, the Federal Republic of Germany and Switzerland.

Detailed presentations were given on the development of telecommunications and electronic data processing, including the development of new and future applications in the 1980s. Mr. James Martin, an authoritative consultant in the computer field, indicated that microelectronics development would continue to be revolutionary beyond 1985. He underlined the difficulty of making predictions for the 1990s, except in general terms by extrapolation from the recent past and present trends in the laboratories of the USA and Japan, because changes during the 1980s would be so far-reaching.

The impact of these technological developments is determinant not only on the computer industry but also on the telecommunication industry, increasing tremendously the capacity of communication systems while reducing cost. The cost of communications and computers will indeed fall dramatically in the next five years. Microelectronics has moreover potential applications in every area of industry, commerce, administration, medicine, farming, etc.

The <u>effect of new developments in employment</u>. Office automation, electronic mail applications, etc., could adversely affect between 30 to 50% of all clerical and management staff in the public and private sectors. The massive introduction of training aids (such as video tapes, television and computers) could reduce the needs for teachers. In the post office, electronic mail services for industry could eliminate many jobs. In industry the spread of automation in production and administration will provide further substantial manpower cuts.

In the discussions, some participants mentioned figures of potential coming unemployment in the Common Market ranging from 6 million in 1979 to 25 million in 1985. It was therefore essential for governments to take a fresh look at the problem and come up with imaginative solutions since little had been done so far to deal with the electronic revolution. Important new investments were required which would create new jobs. Serious consideration should be given to substantially reducing working hours so that more jobs could be found for the unemployed.

Western Europe, according to Mr. Martin, was obliged to follow developments in the USA because otherwise it could not continue to compete in major export markets with North America and Japan. Within the Common Market, member countries had to pick up the challenge and make arrangements so that they would be competitive with each other and collectively create new industries.

Source: ILO.

Conference on micro-electronics at work

Forty-five academics and representatives of governments, trade unions and management met in Zandvoort (Netherlands) from 19 to 22 September 1979 to discuss the socio-economic problems and potentialities of the application of micro-electronics at work. The conference was organised by the European Co-ordination Centre for Research and Documentation in Social Sciences (the Vienna Centre) and was held under the auspices of the Netherlands Ministry of Science Policy.

The Conference brought together specialists from electronics, economics and the social sciences, from both Western and Eastern European countries; it also had contributors from North and South America. Its purpose was not to draw up conclusions but to allow a wide-ranging discussion and debate on the 20 papers which were presented over the course of the 4 days.

Papers were delivered by participants from the Delft University of Technology and the Czechoslovakian Ministry of Technology and Investment on the "state of the art" of micro-electronic technology. They pointed to the current trends in micro-electronics of integrated circuits - a decrease in size and increase in packing density, plus a dramatic decrease in production costs. One problem stressed in the first paper was that the application of micro-electronics will severely test the skills of the engineering profession in the coming years.

On the uses of micro-electronics, participants from the Ontario Communications Education Authority and the US Congress Office of Technology Assessment presented papers indicating the extent to which the convergence of telecommunications network and computing - "telematics" - had already progressed in North America. This convergence could be regarded as a technological revolution and within ten years would have a profound impact on the structure of Western European economies and societies.

A central area of debate was the potential impact of technological change on employment. Whilst some management representatives and academics discounted the likely impact, trade union papers from IG Metall in the Federal Republic of Germany and the British Association of Scientific, Technical and Managerial Staffs (ASTMS) drew attention to the potential for labour displacement arising from automation, made easier and cheaper as a result of micro-electronics (on views of these unions see also S.L.B. 4/79, p. 336 for the FRG and S.L.B. 2/79, p. 122 for the UK). A management representative from the Italian firm Lamborghini also drew attention to the impact that technological change had had in reducing production employment within a single company - Olivetti.

Alternative policies were discussed to effectively harness technological change and avoid unemployment. These included the move to a leisure society, the return to economic growth, the expansion of employment in the personal and social services and the need to thoroughly reform and adapt training and educational policies to a period of accelerating technological change. The need for trade union involvement in the functional design of computer-based systems was also raised in a paper by a participant from the Copenhagen School of Economics.

A participant from the Geneva Centre for Education and International Management raised the question of the impact of micro-electronics on developing countries. One conclusion was that the increase in automated production in industrialised countries as a result of the application of micro-electronics would undermine the cost advantages of developing countries in a number of labour-intensive sectors and lead to a switch of production back to industrialised countries.

In conclusion, the conference agreed that the Vienna Centre should conduct a co-ordination project of research being carried out in different European countries on the social impact of micro-electronics.

Source: Information provided by John Evans, Research Officer, European Trade Union Institute (Brussels).

Banks: a steady advance in automation in next ten years

By 1990, European bank staff directly serving the public will be aided by teller terminals at 60% of counter positions. This is one of the main predictions of a major study by PA Communications and Telecommunications Limited (PACTEL) - a firm of independent management consultants. PACTEL interviewed 270 banks in 13 Western European countries. The study was commissioned by suppliers of bank automation equipment.

Saying that there will be no revolutionary changes in bank automation over the next ten years, the study nevertheless sees a big increase in the number of cash dispensers and automatic tellers - from 6,300 to 50,000 by 1990.

Another finding confirmed the continued use of centralised systems due to heavy investment in such systems and to the centralised nature of management and security.

There is also a cautious attitude to electronic funds transfer systems although there are pilot projects in certain countries, including France and the United Kingdom (see also article on Norway, S.L.B. 3/80, p. 277). Truncation, the method by which banks avoid the expensive and labour-intensive process of physically exchanging and sorting cheques, will also make slow progress in the 1980s.

The study covered building societies and international banks with operations in Europe - but not merchant banks.

Source: Computer Weekly (London), 26 June 1980, p. 16.

INTERNATIONAL

Microelectronics: a double-edged weapon?

"Information technology" is the collective name referring to the convergence of electronics, computing and communications. The unprecedented development of information technology is due to revolutionary changes in the electronic components industry, particularly the silicon-integrated circuit - which contains many interconnected transistors in one "chip" of silicon measuring half a square inch or less. From 10 components per "chip" in 1960 to 64,000 in 1978, forecasts predict 100,000 transistors per "chip" by 1980, increasing to 1 million by 1985.

Nobody really knows what the widespread effects of the new technology will be. According to Dr. Juan Rada, lecturer at the Latin American Faculty of Social Sciences, however, the question today is "whether we are confronted with the advance of technologies that will only provide new elements in an old debate, or whether we are witnessing technological innovations that will mark the end of an era and the beginning of a new one". Opting for the second alternative, he cites four reasons which contribute to the uniqueness of this new information technology: (1) the nature of the new technology which extends and/or replaces human intelligence functions (whereas, in the past, innovations extended or replaced physical strength and dexterity); (2) the unprecedented speed of diffusion and development; (3) the fact that it is based almost exclusively on scientific rather than empirical discoveries by practical men; and (4) its development within a world economy - whereas the initial stages of the Industrial Revolution were limited geographically.

The past five years have seen a staggering reduction in the cost and hence the availability of microelectronic chips. Coupled with the ever-increasing need and demand for information, it seems certain that they will revolutionise many industries.

An OECD study on scientific and technical information needs estimates a growth of about 12.5% per year in the volume of documents produced by the scientific and technical community alone, which should reach a general stock of 120-150 million documents by 1985-87.

Manufacturers of data-processing equipment argue that the proportional increase in the demand for the capture, storage, analysis, processing, retrieval and communication of information is and will be so enormous that even without taking into account decreasing prices, sales will still be profitable. In the United States, for example, it is estimated that 70% of the labour force will be directly or indirectly dependent on some form of data processing by 1985.

The cost reductions, hence the availability of the "chip" is likely to continue. For example, in storage technology, the price of a very powerful computer memory released by Texas Instruments in November 1978, is estimated to fall from 55 dollars per unit in 1978 to 4 dollars by 1985.

Obviously, this has repercussions for social and political decisions. On the positive side there is increased access to vast pools of knowledge; on the negative side there are fears that the high level of unemployment will become more acute and that acceptance of the microelectronic revolution would mean a great upheaval involving the loss of old skills, jobs and workplaces and the organisation of new ones. OECD has already set up an Information, Computer and Communications Working Party to monitor the economic and social implications of the new technology.

The transmission and processing of data has global dimensions - and will affect both developed and developing countries. This is especially true, says Dr. Juan Rada, in the context of the North-South dialogue, the purpose of which is to work towards a New International Economic Order which will eliminate some of the disadvantages of the developing as opposed to industrialised countries. However, in his view, the impact of the new technology will be, by affecting the relative importance of direct labour costs, to diminish the importance of one of the major comparative advantages of developing countries - cheap labour. Moreover, it reinforces the advantages already held by industrialised countries in research and development, design, management, quality control and marketing (not least of these is the substitution of traditional raw materials - the main source of revenue in the developing countries - by development of new materials such as fibre optics, resins and graphites).

Source: New Scientist (London), 8 June 1978.
The OECD Observer (Paris), No. 95, Nov. 1978, pp. 11-16.
Information from Dr. Juan Rada, Specialist in the field of information technology, Latin American Faculty of Social Sciences (Buenos Aires), 1979.

Conference on Human Choice and Computers

The second conference on "Human Choice and Computers" organised by the International Federation for Information Processing (IFIP) took place in Baden (Austria) from 4 to 8 June 1979. The first "HCC" conference had also been held in Austria in 1974.

Over 120 participants were present from throughout the world. Included were people from computer companies such as IBM, Honeywell, Philips and Siemens, from universities and technical institutions concerned with computer science, from government departments and from national and international trade union organisations.

The great variety of subjects covered ranged from the impact of computer technology on employment and working conditions to the effects of computers on the process of making ethical judgements at work and in life generally. There were also widely differing perspectives amongst the participants on how to view the subject: as an academic discipline, or as a subject of immediate and pressing importance requiring concrete political and economic action.

Among the subjects discussed which are of immediate relevance for labour and social policy were: the impact of computerisation on employment in the manufacturing and tertiary sectors; the effects of computers on the quality of working life and job satisfaction; and the need for and problems associated with trade union participation in the design of computer systems. Also, of importance for trade unions, managers and others connected with industrial relations were issues connected with data privacy.

The conference format was divided between formal presentation papers in a plenary session and in-depth study groups on matters of particular interest. Contributions to the plenary session were mainly from academics, although a number of trade union contributions were made: from the Research Institute of the German Confederation of Trade Unions (DGB), from the Austrian white-collar union (GPA) and from the International Federation of Commercial, Clerical and Technical Employees (FIET). Also contributing from the trade union perspective was a member of the Norwegian Computing Centre who is actively involved in developing participative computer systems design methods.

Contributions from the producers were somewhat scarcer, although a useful case study of computer systems design in practice was given by a representative of Philips (Netherlands). The working groups covered subjects such as computers and work; empirical research on the impact of computerisation; and new developments in computer technology.

The conclusions of the group on "computers and work" did not produce a consensus on the likely impact of computerisation on employment. However, despite a wide disparity of views and background within the group, there was significant unanimity on the fact that the employment situation over the next few years would not be easy, and that economic rather than technology policies would have to bear the major task of adjustment. Governments, for example, will need to set priorities for investment decisions, taking into account the full social costs of the employment effects. Trade unions will be right to insist on maintaining workforces unless a reduction is the only possible economic alternative. Above all, however, there was agreement that "participation" in the design and implementation of computer systems was a good thing, although less agreement on who should participate, in what, and at what stage.

As a result of the conference, it is hoped that IFIP will give advice, through its members (the firms and national computer societies to computer specialists and company managers both making and using computers to increase the human input to the work they are doing. This is necessary not just for ethical or humanitarian reasons or because it will increasingly be demanded by trade unions on behalf of employees (although these were recognised as potent reasons), but because in the long term it is necessary for the efficient working of computer systems. As the cost of computers falls, the people who work them become an ever-increasing proportion of the total cost. As a result, the over-all cost to a company, government department or local authority of a dissatisfied and unproductive workforce becomes greater. In the past the criteria on which designers have had to operate in devising computerised systems have neglected to include costs associated with human problems in cost-benefit analysis, with sometimes devastating economic costs to the employer in the longer term.

On the issue of data privacy, the conference heard a detailed commentary on Swedish data protection laws. The importance of data privacy should not be underestimated either by the trade unions or the employers both from the point of view of trade union "blacklists", and general concern about individual privacy. Facilities such as electronic funds transfer,Viewdata (a domestic visual display terminal linked through the telephone system to a computer) and personal computers increase the possibility of people's detailed movements and activities being monitored. Apart from the data itself, the speed with which it can be handled makes possible intrusions into privacy not possible with manual information systems. A particular area of concern is that of "soft data", i.e. details of opinions and comments rather than hard fact, where the origin of completely spurious data from a computer file can give it, in the hands of an uneducated user, the impression of authority.

While the usual job of IFIP is to facilitate technical co-operation amongst computer specialists, this particular conference was an exchange of ideas between specialists and interested parties from other walks of life. The fact that such a conference was held and that IFIP now has a network of technical committees dealing specifically with the human aspect of computing must be considered as an optimistic sign in a world in which the computer is coming to have a bigger and bigger influence.

Source: Information provided by David Cockroft, Secretary for the Industrial Trade Section, FIET (Geneva).

IFAC symposium on criteria for selecting appropriate technologies under different cultural, technical and social conditions

A symposium held in Bari (Italy), 21-23 May 1979, was sponsored by the International Federation of Automatic Control (IFAC) through its two technical committees: one on the social effects of automation and the other on developing countries. The Italian Federation of Scientific and Technical Associations, Centre for Studies of Application of Advanced Technologies and other Italian scientific and training institutions also took an active part in organising this meeting, which was attended by over 100 experts from 20 countries. The developing world was represented by participants from Egypt, Mexico, India, Libya and Saudi Arabia.

The Agenda items included general criteria for the choice of technology; economic factors; organisational factors and diffusion of technical innovations; social aspects of technology transfer; the impact of microelectronics upon the choice of technology.

One of the many papers dealing with general approaches to the choice of the appropriate technology - "Thinking about appropriate technology" by H. Haustein, H. Maier and H. Robinson and developed in the International Institute for Applied Systems Analysis (Vienna) includes systematic socio-economic opportunity analysis with utilisation of formal models for developing tactical decisions.

The practical problems of technology transfer, diffusion and implementation received extensive coverage. For instance the report of H.J. Warnecke, G.F. Pflieger and J. Schmid contains firsthand experience in the developing technology transfer centres. Many social-psychological problems of diffusion and implementation such as "role of the change agent", "resistance to innovations", "the role of the formal and informal structure of the local society" and others were discussed in the reports of H. Schwimann, J. Elizondo and F. Lara Rosano. Another important means of technological and organisational innovation - training - was also covered by several reports.

An important problem for the future development of the socio-technical systems was raised by Dr. Rosenbrock in "The redirection of technology". Pointing out that most present technology was born and developed in a strictly Taylorist spirit, he made an appeal for changing the basic approach to technology design, starting from job content and working conditions.

At a general discussion session the ILO representative briefed the participants on ILO activities, including the International Programme for the Improvement of Working Conditions and Environment (PIACT), and in particular in activities related to the promotion of appropriate technology for the fulfilment of basic needs in the Third World. The ILO publications distributed at the meeting included "Technology to improve working conditions in Asia", an innovative study on the relationship between technology transfer and choice and working conditions in developing countries.

The by now, well-known ideas of E.F. Schumacher concerning intermediate technology, energy savings and so on were reflected in a film entitled "Another way".

Discussions demonstrated that the problem of selection of appropriate technology for the developing countries has in the course of time become more and more complex and that it has a great influence upon social and economic development. Commonsense cannot provide sufficient guidance any longer for evaluating "appropriateness" of the technology. Both fundamental and applied research conducted by multidisciplinary teams of experts are badly needed in order to diminish the harmful effects of the chaotic process of technology transfer nowadays.

Source: ILO.

Club of Rome examines impact of new technologies

The Club of Rome held a meeting at Berlin (West) from 3 to 6 October 1979 on "The coming decade of danger and opportunity". Among the many subjects discussed, three were selected for the preparation of new reports to the Club: the armaments race, the concept of economic value and the impact of micro-electronics on society.

This last question was the subject of a report by Günter Friedrichs, Head of the Automation and Technology Department of the Metalworkers' Union - IG-Metall - of the Federal Republic of Germany.

According to Friedrichs, the effects of technological change in the recent past have nearly always exposed the economically active population to serious short-term hazards (unemployment, loss of qualifications, dehumanisation of working conditions). Moreover, although they may open up in the long run possibilities of creating more wealth and of reducing hours of work, it is always the workers directly or indirectly affected who pay the price for those advantages. It is in this context that the advent of micro-electronics, which Friedrichs considers will be the key technology of the 1980s, must be situated.

The report concentrates on those applications of technology which will have an effect on large numbers of persons. He relies on studies carried out mainly in the Federal Republic of Germany. So far as industrial production is concerned, he shows that the substantial loss of jobs which occurred between 1970 and 1977 in total manufacturing and mining industries, as well as in the sector of office and data-processing machines, suddently accelerated in 1977 - the year when micro-electronics was effectively introduced. In the case of administrative services and of offices, the outlook until 1990 was a bleak one, especially with the introduction of new techniques in word processing, not to mention the possibility of commercialising the automatic typewriter or even the possibility of skipping this phase of development completely. Neither the other subsectors of the services sector nor industry will be able in the medium term to compensate for the job losses caused by micro-electronics; global job losses must be expected in a situation of high general levels of unemployment. Thus, from 1970 to 1977 production in the manufacturing and mining industries in the Federal Republic of Germany increased by 13.5%, while employment dropped by 14.5% (in total: 1.246 million persons) and the volume of employment (number of persons x worked hours) decreased by 21.3%, productivity per hour having increased by 44.3%. The figures for producers of office and data processing machines are even more telling: occupying third place in production growth (48.9%), that sector lost 20,600 jobs (25.8%) while productivity per hour grew by 105.5%.

After reviewing the repercussions of micro-electronics on job qualifications, working conditions and economic centralisation, the report inquires into the long-term contribution of micro-electronics to job creation. The only hope lies in the field of consumption because at the level of investment it economises capital. "In the very long run", Friedrichs concludes, "micro-electronics offer both quality improvements that are so high that additional demand is created and completely new products that definitely will create new jobs. However, there will be an important time lag between the period in which the use of micro-electronics is dominated by process innovation and the period in which product innovations start to dominate".

The serious social effects during the time lag, which might last for five to ten years, might be attenuated by a better distribution of possibilities of employment among jobseekers and by an acceleration of qualitative growth different from that which has been known during the past 30 years.

The Club of Rome did not take a position on the rather pessimistic propositions submitted to it in the Friedrichs report but it confirmed that micro-electronics was a social phenomenon of capital importance and that the Club should study it more thoroughly. That study will go ahead under the direction of Adam Schaff and Günter Friedrichs. It could be ready for the beginning of 1981.

Source: ILO.
Günter Friedrichs: Micro-electronics - A new dimension of technological change and automation. Report to the Club of Rome (West Berlin), Oct. 1979, 27 pp.

Congress debates New Information Order

The concept of a New World Information Order still has a long way to go before becoming a reality. Delegations to the World Conference on Transborder Data Flow Policies, organised in Rome from 22 to 26 June 1980 by the Intergovernmental Bureau of Informatics (IBI), arrived at no consensus on how to constitute a "new order" of international and national information systems.

Although universal access to sophisticated information services and technologies, free from domination of any one country or group of countries remains an ideal, it will not, it seems, be easily achieved.

The industrialised nations generally argued for free flow of information. Several Third World delegates among the 60 nations in attendance, however, claimed that national, political, economic, technological and cultural considerations mandate at least some control over information flowing into and out of their countries.

Concern was also voiced over the dominance of the industrialised nations in international information services and a fear that developing countries may become mere client States to the large multinational corporations that control existing information technology.

Source: Computerworld (Framingham, Mass.), 23 June 1980, p. 2.

Technical change and economic policy: an OECD study

Why, in a recession, expect a recovery in demand to stimulate technical change, when the inadequate rate of technical change may itself be a source of slack demand?

This question very briefly summarises the study made by a group of industrialists and university specialists under the auspices of the OECD Committee for Scientific and Technological Policy, in order to identify the impact of the economic and social changes of the past decade on research and innovation and to examine the circumstances in which such activities can help industrialised countries to overcome present difficulties.

In the opinion of the experts these difficulties, although considerably aggravated by external events (particularly the oil crisis), go back to at least the late 1960s, when four groups of changes combined to create an entirely new economic and social context:

- the slowdown in economic growth accompanied by persistent unemployment and inflation;
- the new distribution of economic power within the OECD area where the role of "locomotive" no longer belongs exclusively to the United States but is

shared with Japan and Western Europe (especially the Federal Republic of Germany);
- the determination of oil prices by essentially political considerations;
- the emergence of new social values and aspirations (importance attached to protecting the environment, changes in attitudes to work, and a more critical assessment of science and technology).

Competition from the industrialising countries calls for the OECD countries to change the composition of their output, and their intellectual capital (obviously linked to research and development capacity) is their major asset.

However, the rate of innovation has slowed down in most of these countries, and research - the cost of which has risen steadily - has been oriented to short-term, low-risk projects. With the exception of electronics and bio-engineering, where breakthroughs are constantly being made (computer/telecommunications, word processing and automation) many sectors until now regarded as strongly innovative (pharmaceutical industry, pesticides, etc.) are marking time owing to the cost of environmental safety and protection. Since 1973, in particular, there has been a persistent slow-down in labour productivity. The composition of the workforce (which is tending to shift from manufacturing to lower productivity sectors) and measures to combat inflation have contributed to this phenomenon, which is further accentuated by the fact that Gross National Product as measured at present does not set any value on environmental quality, and the resources that are devoted to it show up in the form of a decline in measured productivity growth.

Whether it is a cause or an effect, the slowdown in technical progress - and consequently in labour productivity - merely aggravates the difficulties faced by industrialised countries in their fight against inflation and unemployment.

Employment, which increased in industry at the expense of agriculture during the 1950s, is now tending to decline in industry; nor is it certain whether the services sector will be able to make up for this decline. Moreover, the bias towards capital-using technologies (micro-electronics) may have adverse effects on employment in the medium term.

In this context, technical innovation, far from being peripheral, becomes essential and the rate and direction of technical change are at the heart of economic policy options.

This conclusion led the group of experts to emphasise the importance of three objectives:

- maintaining and improving the innovative capacity, not only in manufacturing and the services but also at the level of fundamental research, which must be shielded from the consequences of recession;

- sustaining a higher rate of technical advance and productivity increase by laying stress on fundamental technologies (materials resistance, corrosion control, etc.) which may have wide application in a number of essential sectors. The door should be kept open for developing alternative technologies in order to avoid being caught out by political or technological surprises (such as in the energy crisis);
- promoting social innovation and technologies: this calls for special support from the public authorities since the organisation of demand in this field is less clear (transport, health, urban development, etc.).

The experts conclude that, if the health of the innovative system is to be restored and a higher rate of technical change is to be accepted, there should in the first place be a satisfactory balance between the generation of new employment and the loss of old jobs, and change must be perceived to improve the quality of life. This calls for the establishment of new sorts of relationships between policy makers and scientists, engineers, trade unions, consumers' organisations and representatives of the public.

Source: OECD: Technical change and economic policy: science and technology in the new socio-economic context (Paris), 1980, 133 pp.
The OECD Observer (Paris), No. 84, May 1980, pp. 16-22.

OECD research agenda for transborder data flow

A major international study of the economic implications of transborder data flow is planned by the Organisation for Economic Co-operation and Development (OECD).

The OECD's expert group on transborder data flow will be presented with a number of proposals from OECD member countries. In addition, the expert group will be asked to study the growth of the international use of on-line computer services to determine the market dimensions of those services and their effect on the growth and structure of the computer services industry.

Source: Computerworld (Framingham, Mass.), 19 Jan. 1981, p.18.

PRIVACY PROTECTION

NORWAY

Curbs imposed on abuses of privacy

Legislation which came into effect on 1 July 1980 gives individuals the right to see all the facts about themselves that are fed into public and private data banks (the bulk of the legislation took effect on 1 January 1980).

The new legislation also stipulates that there must be a reasonable relationship between the need for a data bank and the information it contains. It also imposes stringent restrictions on such information as religion, race and political beliefs.

However, it is not a case of complete openess. Medical information which would be harmful to a patient is not divulged, and national and police security registers will be a closed book to most people.

Source: Norway: Lov om personregistre m.m., Law No. 48 of 9 June 1979, Norsk Lovtidend (Official Gazette), 1978, pp. 402-414.
Norinform (Oslo), No. 25, 8 July 1980, p. 1.

UNITED KINGDOM

Computers and privacy

The report (Cmnd 7341) of the Committee on Data Protection set up to advise the Government on machinery to safeguard the privacy of personal information held on computers was published at the end of 1978. The report discusses in detail the problems of particular fields of personal data handling, including employment records. The Committee recommends legislation to define the principles governing data protection and the establishment of a Data Protection Authority (DPA) to implement them.

The Government will hear the views of Parliament, computer users and others before taking further action.

Source: Survey of Current Affairs (London), HMSO, No. 1, Jan. 1979, p. 16.

Government urged to issue privacy statement

The National Computer Users' Forum (NCUF - an association of 25 user groups) has asked the Government for a definitive statement of government intention with regard to the issue of computers and privacy. This would allow computer using organisations to plan ahead with less uncertainty.

The NCUF has already recommended that any "codes of practice" be given a voluntary trial period to illustrate their pertinence to the privacy question before any attempt is made to incorporate them into legislation.

The Government is considering international developments in this field, particularly on-going discussions at the Council of Europe and OECD, as well as the views of national interest groups - ranging from support for stringent safeguards to concern that controls should not add to industrial costs in a competitive market.

Source: Computerworld (Framingham, Mass.), 23 June 1980.

UNITED STATES

Government protects individual privacy against abusive use of computerised data

On 2 April 1979, the United States President submitted to Congress a special message on the protection of individual privacy, stating that while "modern information systems are essential to our economy ... they can be misused to create a dangerously intrusive society".

At the same time the following three bills were submitted to Congress to establish a framework to prevent abuse of individual privacy.

The Privacy of Medical Information Act would ensure confidentiality for information maintained by medical institutions. An individual would be permitted to see his own medical records and to object to information that was not accurate, timely or relevant. The Fair Financial Information Practices Act would provide full information protection concerning banking and insurance records, while the Privacy of Research Records Act would ensure that material collected for research purposes could not be used to adversely affect the individual.

No legislation would be proposed on employment records; instead, employers should be asked to establish voluntary policies to protect their employees' privacy. However, the Administration supported a bill (No. S.854), already before Congress, to limit the use of lie detectors in private employment.

Parallel to these initiatives, guidelines have been drawn up to protect personal records kept by federal agencies. The Federal Government holds almost four billion records on individuals, stored in thousands of computers. Federally-funded projects have substantial additional files. Modern technology, however, makes it possible to turn this material into a "dangerous surveillance system". Government data on private individuals will, under the guidelines, come under stricter procedural restrictions.

The enormous increase in personal data records in the United States has been matched in other industrialised countries. The United States is co-operating with other governments in several international organisations to develop principles to protect personal data crossing international borders.

Source: US Code Congressional and Administrative News (Washington, D.C.), No. 3, May 1979, pp. 522-527.

EUROPE

Council of Europe adopts convention on data protection

On 17 September 1980, the Committee of Ministers of the Council of Europe adopted a Convention for the protection of individuals with regard to automatic processing of personal data. (This convention was discussed in its draft form in S.L.B. 1/80, p. 13.)

The convention will be open for signature by member States at the end of January 1981 and will come into force three complete months after the date on which five member States have expressed their consent to be bound by the convention by ratification, acceptance or approval. The convention is also open to non-member States and to non-European States.

Source: Secretariat-General of the Council of Europe: Convention for the protection of individuals with regard to automatic processing of personal data. Provisional edition (Strasbourg), Oct. 1980, 12 pp. (mimeo.).

Privacy and researchers' need for access to data

According to the European Science Foundation, the implementation of data protection laws has led, in an increasing number of cases, to serious restriction on access to data for research purposes. The ESF is particularly concerned about the destruction of data by public authorities once its original purpose has been served.

While fully accepting the need for legislation on this subject, the ESF believes that such laws can achieve their aims while at the same time ensuring access to personal data for research purposes. The ESF has recently published a statement which sets out a number of principles and guidelines by which these two objectives can be met.

According to the ESF, data protection legislation must, in order to fulfil its task, which is to guarantee the respect of privacy, cover all uses of personal data and therefore include its use for research purposes. Any use of personal data for research purposes should be granted by law or by the informed consent of the individual unless the individuals concerned are not identifiable. In addition, the ESF advocates the development of techniques to secure anonymity of data and the establishment of legally recognised procedures for using certain type of data collected by state agencies for non-research purposes.

Any decision to destroy personal data held by the public authorities, says the ESF, should only be taken after consideration of their possible future use for research. However, the individual's right to challenge personal data should be limited to those research projects where it is intended that the data be used in an identifiable form.

The ESF, which was set up in 1974, is an international, non-governmental organisation grouping 47 research institutes from 18 countries.

Source: ESF: Statement concerning the protection of privacy and the use of personal data for research, adopted by the Assembly of the ESF on 12 November 1980 (Strasbourg), 13 pp.

INTERNATIONAL

Computers and privacy

Reflecting the sense of concern caused by more and more information becoming available through computerised data bases containing personal information, a draft international convention on computers and privacy is currently being examined by the Council of Europe's steering committee on legal affairs. The plenary session on the text is scheduled for 23 to 27 June 1980. Following approval, the text will be submitted for adoption to the Council of Europe's decision-making body, the Committee of Ministers.

Apart from the countries that make up the Council of Europe, five non-member countries have followed the negotiations as observers: Australia, Canada, Finland, Japan and the United States.

The draft convention is based on a "common code" of principles which place obligations on the users and confer rights to the data subjects. Users should ensure correctness, fairness and security when handling personal information. Data subjects would be allowed to know and, where necessary, to challenge the information stored about them.

Special provisions will be included in the convention to help nationals in one country to exercise their rights with regard to data files in another country.

The draftsmen of the convention have encountered two problems. One is a legal puzzle: if an automated data file in country A, owned by a company in country B, processes information on a data subject in country C, whose law applies? This question has been referred for further study.

The second concerns trans-border data flows between countries having a different level of data protection. The general rule, according to the steering committee, should be no barriers between contracting parties.

With respect to the convention, a special role is reserved for the European Economic Community (EEC). In May 1979 the European Parliament asked the EEC Commission to take steps to deal with data protection. The Commission has decided to await the conclusion of the Council of Europe's convention. The convention will also be open to non-European countries. The text has been drafted in close co-operation with the Organisation for Economic Co-operation and Development (OECD) which is also drawing up privacy guidelines. Moreover, on 5 June 1979, in a resolution on the "rights of the individual in the face of data processing", the European Parliament called on member States to co-ordinate their efforts in all international forums and, once the Council of Europe convention has been signed, to work for the access to that convention of the greatest number of third countries subject to reciprocity. The accompanying recommendations stressed the right of individuals to be fully informed on all measures involving the recording, storage or transmission to third parties of data relating to them, and to have inaccurate or incorrect data subsequently corrected.

Legislation in the various countries in Western Europe is patchy. So far, only Austria, France, Denmark, the Federal Republic of Germany, Luxembourg and Norway have passed special laws introducing privacy safeguards for computerised data files. Sweden is considering draft legislation and the question is receiving serious consideration in several other countries.

In the United States (see S.L.B. 3/79, p. 236), proposed legislation covers the confidentiality of medical and research records, personal information kept by government services and financial information. In addition, the Department of Labor started a public inquiry in January 1980 into the protection of privacy rights in the workplace. The inquiry will examine the extent to which employers observe the recommendations issued in 1977 by the Privacy Protection Study Commission. However, American companies are especially sceptical about data protection legislation and see it as a constraint on corporate data collection, dissemination and use.

Source: European Communities: Official Journal, Information and Notices, C.140, Vol. 22, 5 June 1979, pp. 34-38.
United States Department of Labor: News, 7 Jan. 1980.
Computerworld (Framingham, Mass.), 15 Oct. 1979, 21 and 28 Jan. 1980.

Computers and privacy in the ILO

The collision between the legitimate needs of institutions for information about people with the rights of those people to be assured that such information is not used incorrectly, is the hub of privacy.

The ILO dealt with the problem by setting up a joint working group to re-examine practices for recording, maintaining and using personal data, as well as for the rectification and erasure of such data. The ILO Staff Union had called for the establishment of such a group at its annual general meeting of 20 September 1979.

The first recommendations of the group covering access to information have been implemented. All ILO officials, whether at headquarters or passing through headquarters from assignments elsewhere, are now entitled to see all the facts about themselves that are fed into the personnel data bank. They can also ask for incorrect information to be changed or removed. Steps have also been taken to ensure that no unauthorised information is communicated to a third party.

Source: Access to personal information on the computer, ILO Circular (Geneva), Series 6, No. 190, 25 Nov. 1980.
Resolution on personal information, ILO Staff Union Annual General Meeting (Geneva), 20 Sept. 1979.

LABOUR RELATIONS AND COLLECTIVE AGREEMENTS

AUSTRALIA

Dispute over introduction of new technology settled in telegraph and telephone services

At the end of August 1978, the Australian Telecomunications Employees' Association (ATEA), endorsed a settlement with Telecom Australia which ended a month of industrial action which had crippled telephone and telex services. The dispute had its roots in the introduction of new technology and changes in technical staff structure concerned with maintenance work.

Some of the union's main concerns were that the introduction of computerised telephone exchanges could result in the downgrading of certain jobs, reduced promotion opportunities, staff cuts and extensive redeployment.

The settlement which was reached in the Commonwealth Conciliation and Arbitration Commission contained these basic elements:

1. Despite a review of the technical staff structure, continued job status is guaranteed for technicians and any restructuring will provide reasonable promotion opportunities. Areas of disagreement will be referred to the Commission.

2. Both sides agreed to a trial period of two years to examine the job levels required to service the new equipment. This trial period will be reviewed by two independent experts (one nominated by the unions and the other by Telecom) with their findings being the subject of a public debate before the Commission. Factors relevant to the trial period and investigation include: efficiency of operation; standard of service achieved; job satisfaction; career opportunities; maintenance of technical standards and retention of expertise; the public interest. No industrial action will be entered into which might prevent a fair trial period.

3. Full wages were guaranteed to nearly all ATEA members who were suspended for refusing to perform certain jobs associated with the new technology. Only those who refused to carry out duties at their regular work post will not be paid. Those who refused to work in areas away from their normal duty station will not suffer loss of wages.

In an attempt to prevent a similar dispute arising in the future, the agreement provides for consultation to take place between Telecom and the ATEA before technology is introduced and if the parties cannot agree, the matter is to be referred to the Commission.

The ATEA remains concerned about the potential of

computerised technology for deskilling work and eliminating jobs. The ATEA believes that a broadly based public inquiry should be established to examine all aspects of the introduction of new technology. The President of the Australian Council of Trade Unions has called for such an inquiry.

Source: ATEA: Teletechnician (Sydney), Sep. 1978.
PTTI News (Geneva), No. 9, Sep. 1978, pp. 1-2.

Special pay increment to account for the additional demands of work with VDU's

An arbitration agreement recently reached by the Federated Clerks' Union of Australia (FCUA) and the Australian Civil Servants' Association concerns proficiency and functions of text processors and supervisory staff. It provides for about 8 per cent pay rises on the grounds of the additional demands made on typists who have to work with "third generation" machines with visual displays. At the same time, the arbitrator stated that text processing results in increased job satisfaction, since this work is "creative and demands both imagination and analytical discernment to a degree which far exceeds that required by more routine clerical work".

Source: ICFTU: International Trade Union News (Brussels), ITN/9, 1 May 1980, p. 7.

AUSTRIA

Collective agreement on text processing in the printing sector

Following several years' negotiation, the Journalists Section of the Arts, Media and Liberal Professions Union, the Printing Union and the General Workers Union signed an agreement on 11 May 1981 with the graphic industries and newspaper owners organisations on the introduction of an integrated text-processing system (ITS) in the composition of dailies and weeklies.

Under the agreement new electronic technologies will be introduced in the production process of all newspapers within the next ten years, with consequent redundancies.

All employees affected will be protected by a two-year job security clause against dismissal, rising to ten years for typesetters, compositors, proof-readers and related draftsmen if they have five years' service.

Newspapers introducing ITS shall inform and consult works councils on planned changes as soon as possible.

Normally, when any of the contracting parties is informed about the planned introduction of ITS it shall inform the others.

For eight years after the introduction of ITS, the employers shall announce all vacancies and new jobs in the company. Priority for these jobs will be given to redundant workers; if they do not have the appropriate qualifications, the employer shall provide the necessary retraining.

Third parties (e.g. advertising companies, news-agencies and other newspapers who are not an integral part of the company) shall not enter data directly through the new ITS for five years following its introduction.

Job protection for workers aged 50, with at least five years of service, will be guaranteed until they reach the earliest possible retirement age.

In cases of transfer to a lower-paid job, workers may request the difference between their former and present remuneration. The final amount will be determined together with the works council.

Where necessary, the employer shall provide retraining which must be accepted by the workers, during which they shall retain their former remuneration. A retrained worker must be capable of operating the ITS after four weeks' practice following the course. If, after two courses, the employer finds the worker's performance inadequate, the case will be submitted to arbitration.

Mobility assistance amounting to a maximum of one-and-a-half year's remuneration will be given to employees for whom no acceptable job is found and who resign. They will also be entitled to normal severance pay as if they were dismissed.

Editors have to check all articles before they are forwarded to the technical processing units.

Proficiency in operating terminals cannot be used to assess journalistic performance. VDUs shall normally be used by journalists for proofreading, editing and revision of texts. They may be asked to enter first drafts only if they previously used to type articles. Entry of typographical instructions is, however, considered to be part of the journalist's job.

Journalists may refuse to work on VDUs for eight years following ITS introduction (until retirement for those aged 50 with five years of service) without affecting their terms of employment.

Source: Federation of Austrian newspaper owners: Killekt-ivvertrag über di Einführung von integrierten Texterfassungssystemen bei Tages und Wochenzeitungen (ITS-Vertrag), effective as of 11 May 1981 (Vienna), 15 pp.

DENMARK

Central agreement on technology

The introduction of new technology in Danish industry is now governed by an agreement under which worker representatives must be informed in advance about changes "of major significance" and their likely impact on jobs.

Under the agreement, effective 1 March 1981, both the Employers' Confederation (DA) and the Federation of Trade Unions (LO) confirm their commitment to new technology as a positive method of improving competitiveness, boosting employment, and improving job satisfaction and the working environment.

The agreement provides for permanent joint technology committees in which all aspects of proposed new technology can be discussed - technical, economic, staffing and environmental. The committees have joint decision-making powers over basic principles for employee training, the use of personal data, and retraining for those whose jobs are affected by the changes.

Where the introduction of new technology involves a loss of jobs, undertakings must attempt to find alternative employment. If declared redundant, workers will be entitled to reasonable time off to attend retraining courses of not longer than 2 weeks. The employer will supplement any state financial assistance to ensure that the cost of the course is covered and that the employee suffers no loss of income.

The new agreement runs until 1984 and supplements the 1970 DA/LO agreement on "co-operation committees" which established a framework for joint discussions and decision-making at plant level and outside the collective bargaining framework. Realising that technology will bring sweeping changes into the workplace, the LO sees the agreement as responding to the need for employees to be kept fully informed about management's plans for the introduction of new technology.

Source: LO Bladet (Copenhagen), 27 Feb. 1981, pp. 8-9.

Printworkers back in print

A dispute involving about 11,000 printworkers kept the majority of Danish papers out of print for 11 weeks during spring and early summer 1981. The dispute was finally settled in the middle of June and the striking or locked out printers resumed work on the 15th of that month.

The basic reasons for the dispute were wages, job security and a union demand for veto on technological change. The cause of the dispute was the collapse of negotiations for a new agreement on the statutory expiration of the previous agreement within the framework of the general national agreement negotiated between the Danish Federation of Trade Unions (LO) and the Danish Employers' Confederation (DA).

Negotiations were carried out under four headings: wages, skill levels and demarcations, social matters and technology. It was mainly wages and technology that caused the problem. As far as wages were concerned it was probably mainly a tactical move on the part of the printing workers in order to get concessions on technology. Up till the last moment security of employment remained the main bone of contention.

The demands of the printworkers were not met. All major points that the employers' trade organisations were willing to concede were blocked by the DA's central position on principles.

Final agreement was reached by the intervention of the LO in negotiations with the DA. This included wage increases of about 8% in 1981-82. No satisfactory agreement was reached on technological change as had been the case in other industries negotiating under the general national LO-DA framework agreement (see S.L.B. 2/81, p. 133).

Source: Politiken (Copenhagen), 16 June 1981.
Weekend Avisen (Copenhagen), 19 June 1981.
Berlingske Aften (Copenhagen), 19 June 1981.
Le Gutenberg (Lausanne), 14 May 1981.

FRANCE

Metal trades employers launch social data bank

Faced with a jungle of legal and contractual provisions and increasing difficulty in handling an ever-growing mass of legal and social data, the Federation of Metallurgical and Mining Industries (UIMM), the employers' union to which some 15,000 undertakings are affiliated, has set up its own legal and social data bank - the Automatic Document Handling Transaction System (SIGAD) - which has been operating since April 1980.

The data bank, containing some 90,000 documents, is designed to supply the user with specific and concise documentary material both for rapid consultation and for more detailed study in connection with litigation.

The source material stored will be that used regularly by legal experts in the social field: Labour and Social Security Codes, parliamentary replies by ministers, ministerial circulars, official civil service commentaries, full case law of the Social Chamber of the Court of Appeal since 1 January 1976, selected case law for other courts, all collective agreements and national agreements covering the metal trades.

Access to the data bank is reserved in principle to employers' organisations and affiliated undertakings against payment of a subscription and an access fee. Users must buy or rent terminals and connect to the transmission network.

At the press launching the UIMM representative said that the possibility of communicating data to the trade unions was not ruled out.

Source: IUMM: SIGAD, press handout (Paris), Mar. 1980.
Le Monde (Paris), 29 Mar. 1980.

Employee access to information on data processing systems

It is clear from the parliamentary debates in connection with the data processing and privacy legislation (voted on 6 January 1978) that no real dialogue has yet developed between labour and management on proposals for the introduction of data processing systems.

The National Agency for the Improvement of Working Conditions (ANACT) and the Information Technology Commission have now completed an in-depth review of the problem with a view to providing guidelines for amending current legislation and regulations.

The review contains details of the legal machinery governing workers' access to information, followed by practices in eight undertakings in the public and private sectors. Following the same pattern, it also makes a comparative study of the law and practice in Denmark, Federal Republic of Germany, Italy, United Kingdom, United States and, in greater detail, Norway and Sweden. (The right of employees to be kept fully informed about the introduction of computer technology systems in these countries has been analysed in various issues of the Bulletin; see, for example, S.L.B. 2/79, p. 121, 4/79, p. 326, 2/80, pp. 147 and 176, 3/80 p. 280.)

In all of the undertakings studied, information on computerisation projects appears to circulate through official channels, i.e. those provided for by the Labour Code (mainly works councils) or by civil service regulations (joint technical committees), as well as through direct hierarchical channels on the initiative of either the management (information meetings or bulletins, company news papers, etc.), the trade unions or certain categories of staff (e.g. computer staff).

According to the report, fairly rapid and unbiased information is normally supplied through the official channels in both the public and the private sector when it is intended for the management of the undertaking or particular service concerned. However, relevant information tends to be much slower in reaching bodies representing the staff of a factory, business or department, who may nevertheless be directly affected by the technological developments. At this level, information tends to be supplied "after the event", dealing with choices already made or with their consequences, rather than with the possible options.

Information supplied through the parallel circuit, which operates alongside this first channel, is more direct and explicit but is likely to be less objective. Nevertheless it serves as an effective means of double-checking data supplied through official channels, particularly in the public sector where powerful and dynamic trade unions are in a position to supply their own additional information and even to contradict that provided officially.

The proposed modifications aim essentially at enabling statutory bodies to react sooner, and at every stage of a project.

In particular, the report recommends that the statutes of the works councils should contain a clause making it compulsory for employers planning to introduce computerised operations to: explain to the council the future impact of the planned measures on employment, work organisation and working conditions, staff training, etc.; attach to their annual report to the council a progress report on computerisation showing the amount of investments, subcontracts, etc.; organise once a year a meeting with the council so that it may state its views on any computerisation plans; establish an ad hoc committee including, besides council members and representatives of the management as well as of the services affecting by computerisation; allow the council to call in outside experts at the undertaking's expense (up to 1% at most of the cost of the computeristion programme); and to prepare a list of social commitments before adopting any computerisation programme.

Similar commitments are recommended for the public sector.

In branches or undertakings where there is no consultation machinery of this kind and where the collective agreement is the only means through which workers can influence computerisation projects, it is proposed that it be made compulsory to include a clause on "the conditions of employment and work of persons assigned to computer-related jobs" together with a list of the social aspects subject to compulsory negotiations with the trade unions.

Information, as illustrated by the case studies forming the second half of the report (for the private sector a mail order firm, a press agency and two industrial undertakings; and for the public sector the national electricity company, the postal and telecommunications service, hospitals, a nationalised insurance company (see following article), a pension fund and a town) confirm the dual nature of information channels and show that the standard and effectiveness of information depend in practice on the interaction of the two systems. In any efforts to improve communication with the workforce such duality must be retained if not broadened; this should be done both by strengthening the institutional machinery and by according a certain legitimacy to procedures developed by employees and trade unions. The possibility of calling in outside experts, which seems to offer a means of guaranteeing the independence of the information process, should be encouraged.

Much ground still has to be covered before workers are not only informed about but are also consulted or even involved in the planning and execution of computerisation projects, though some larger companies are taking tentative steps in this direction. So far, however, such practices are still at the experimental stage and the decentralisation of responsibilities hardly extends below the supervisory staff who already play a key part in setting up computerised systems. Such decentralisation should maintain this pluralism and be extended to the workers themselves, as the direct operators of the future systems.

Source: Association pour la recherce sur l'emploi des techniques (ARETE): L'information des salariés sur les projets informatiques, dossier d'étude. Mission informatique, ANACT (Paris), Jan. 1980, 355 pp.

Computer technology in the insurance sector: union wins safeguards

The use of computers in insurance companies, which first began in 1960 and then was given a further boost by legislation of 17 January 1968 streamlining the national insurance sector, has had far-reaching effects on the volume of employment and the quality of work.

All insurance jobs have gradually been affected by an organisation of work centred on computerised data entry, and the process of deskilling that has changed the job content of many related activities: while filing clerks, cashiers and typists have been the hardest hit, the higher-level responsibilities of underwriters and department managers have also been affected by the introduction of inter-company agreements greatly streamlining the legal criteria for settlements, doing away with the "negotiation" component and enabling inter-company offsetting to be made on a strictly accounting basis simply by cross-checking the respective files. Insurance policies have also been streamlined for the purpose of data processing, by introducing standard guarantees limiting the interpretation of their clauses to a minimum.

Furthermore, the reorganisation (mergers and regional decentralisation) that was both required for and facilitated by computerisation has led to staff retrenchments, through natural attrition without, for the time being, resulting in dismissals.

At Assurances générales, the third largest (nationalised) insurance company in France, the CFDT (the majority trade union in the company) has reacted in two ways to these developments. First, it insisted that a job security clause be introduced in the works agreement of 1973 providing that none of the measures for decentralisation and computerisation then envisaged would lead to redundancies or compulsory early retirement.

Second, it insisted that the various projects for computerisation be referred to the central works council for consideration as part of its legally recognised functions. As a result, in March of each year a meeting of the council is devoted exclusively to computer-related problems. This enables workers' representatives to examine proposed investments (cost of data processing equipment in relation to the company's turnover or payroll, choice of equipment, etc.) and assess their impact on the workforce, working conditions (e.g. ergonomic or medical aspects of work with screens) and on work organisation (content and distribution of tasks, utilisation of gains in productivity, etc.).

This experience has underlined the importance of fully informing the workforce, although the highly technical and specialised aspects of the data supplied are still an obstacle. The trade union considers that the council should be able to call in outside experts to examine the company's accounts.

Source: CFDT Union of Executives and Engineers: "L'action d'un comité d'entreprise pour le contrôle de l'informatique", in Cadres (Paris), Nov.-Dec. 1979, pp. 58-60.

FEDERAL REPUBLIC OF GERMANY

Six-weeks' annual paid leave and guarantees in case of rationalisation in chemicals, paper processing, printing and insurance

The basic agreement for the chemical industry, concluded on 24 March 1979 and effective until the end of 1984, covers some 600,000 workers. It provides for a gradual increase in paid annual leave, reaching six weeks for all workers by 1984 (30 days instead of between 22 and 27 as at present). Workers aged 40 or over will be entitled, as from 1983 on, to six weeks' leave. Similarly, daily holiday bonuses will rise by annual stages from DM25 in 1979 (DM23 previously) to DM30 in 1984 (1 US dollar = DM1.84). Holiday bonuses for young workers and apprentices aged 18 or over will gradually increase from DM400 in 1979 to DM660 in 1984.

The working week for those on alternate shifts will decrease from 42 to 40 hours in two stages (to 41 hours on 1 May 1980 and to 40 hours on 1 May 1982). Loss of earnings will be compensated under transitional regulations.

Redundancy pay for workers laid off as a result of rationalisation has been increased to: one month's wages for workers over 40 years of age with 10 years' service and to six months' wages for those over 60 years with 25 years' service.

Workers undergoing company retraining as a result of rationalisation measures will continue to receive from their employer their former wage for up to six months.

Guaranteed earnings in the event of rationalisation entailing downgrading are now the same for manual and non-manual workers: 100% of former pay during nine (instead of six) months. This also applies to workers over 50 years of age with at least ten years of service who are transferred to another work post.

Entitlement to guaranteed payments from the chemical industry's wage compensation fund in the event of unemployment has been extended: a worker now needs only 6 years' employment in the sector whereas previously 4 years' service in the same enterprise was required.

The basic agreement for the printing sector, concluded on 12 April 1979, provides that the normal 40-hour working week will be spread over five days. Works rules providing for a 6-day week will expire on 31 December 1982. In exceptional cases where the 5-day week cannot yet be applied on that date, a special works agreement will have to be concluded.

Paid annual leave will gradually be increased to six weeks (30 days) for all workers by 1983, though workers aged 35 or over will be entitled to the six weeks beginning in 1979.

Shift workers continuously rotating shifts of equal length or permanently on night shift will receive, beginning on 1 January 1980, compensatory leave equal to one shift for every 6 months so worked and, beginning on 1 January 1982, leave equal to one shift for every 4 months so worked, without loss of wages.

Workers aged 58 or over at the beginning of the calendar year will receive, as from 1 January 1980, one day extra leave for every calendar year and, as from 1 January 1982, two days' leave, without loss of wages.

The basic agreement concluded on 6 July 1979 in the paper processing industry is effective until 12 December 1984. It provides for a gradual increase of paid annual leave to six weeks (30 days) by 1984 for the 55,000 workers in the sector; workers aged 29 or over will be entitled to the six weeks beginning in 1982. Shift workers will receive an additional two days of leave (the first in 1979 and the second in 1983). Also, the daily holiday bonus will gradually increase from DM24 in 1979 to DM30 by 1984.

The agreement also gives better income protection to workers downgraded because of rationalisation measures. Compensation varies according to the age and length of service of the worker affected. For workers over 55 years of age (60 previously) with 25 years' service, such compensation can last for a maximum period of 18 months.

The agreement concluded in May 1979 for employees in the private insurance companies of the DBV Group (Deutchen Beamten-Versicherungs-Gruppe) provides for a general extension of six weeks' paid annual leave by 1 January 1980.

A 39-hour week will also be introduced by reducing work on Friday by one hour.

Source: IG Chemie-Papier-Keramik: (i) Mantel-Tarifvertrag für Gewerbliche Arbeitnehmer und Angestellte der Chemischen Industrie, 1979, 63 pp.; (ii) Presse dienst (Hanover): communiqués, 23 Mar. and 9 July 1979.
Bundesverband Druck and IG Druck und Papier: Tarifvereinbarung (Frankfurt), 12 Apr. 1979, 3 pp.
IG Druck und Papier (Stuttgart) (Weekly paper), No. 13, June 1979.

Union control of personnel data: IG Metall and VW agreement

The metalworkers' union, IG Metall, has recently voiced its demands for full co-determination rights of works councils in the nomination and termination of persons who are entrusted with data protection within companies (i.e. data protection officers). Reacting to a Bill now under discussion in the Federal Republic of Germany (FRG) to increase protection against dismissal of data protection officers, IG Metall contended the proposed legislation would only apply to employees who are appointed as data protection officers on a full-time basis. Since so far there are no full-time officers in the metal trades, the Bill will not improve the present situation.

The union mentions two major areas for which computerised personnel information systems are used and can be misused:

- analysis of workplace structure to identify trends and impact on staff of planned rationalisation measures;
- cost assessment of the undertaking's production requirements considering "flexible manpower input".

The determination of data and programmes for the assessment of these rationaisation measures is of prime importance to the union. For instance: What personnel data are stored? For what purpose? How do decisions taken on the basis of this data affect the workers? And what information on these matters is given to the workers and their representatives?

VW data protection agreement

To protect workers against misuse of personnel records, an agreement was concluded between the management and works council of the car manufacturer Volkswagen in Wolfsburg on 16 July 1979. It covers all company employees and their families, and concerns collection, storage, changing, deleting or transforming personnel records (including linkage with other data) whether or not in computerised data bases.

Under the agreement the management undertakes to provide the works council in good time with a comprehensive description of its personnel information system PEDATIS before expanding it or before introducing any other similar system, as well as to inform the council about the implementation of such system.

Employers are to use personnel data only for purposes related to the employment relationship.

If the processing of the data involves its transmission to a third person, agreement of the works council is required.

The management and the works council undertake to protect the staff and their families from the dangers and negative effects resulting from the processing of personnel records and to avoid any illegal or inadmissible action.

A special list is established by a separate agreement, indicating all data bases containing personnel records, including PEDATIS. The list, which will be regularly updated, can be consulted in the VW personnel department and in the offices of the works council.

VW employees may ask for information on the records that concern them. If the data is processed by computer, the person concerned may request information on the person or body to whom the data is regularly forwarded. This information is provided free of charge.

The works council rights of consultation and information are guaranteed with regard to the appointment and termination of the data protection officer.

The officer informs the works council on behalf of the management about his activities.

The agreement sets up a Data Protection Committee, consisting of: 7 members nominated by the works council, the data protection officer of VW, and representatives of the personnel department dealing with data protection systems analysis and social welfare.

The Committee deals with issues related to: the implementation of the agreement; its eventual amendment; the implementation of the 1977 federal law on data protection in the company; the description of and prior information on PEDATIS and similar systems; and data protection measures.

Source: VW: Betriebsvereinbarung No. 4/79: Personenbezogener Datenschutz (Wolfsburg), 16 July 1979, 4 pp. IG Metall: Der Gewerkschafter (Frankfurt-on-Main), No. 12, Dec. 1979, pp. 6-11, and Metall Pressedienst, 5 Mar. 1980, 2 pp.

Agreement on operating standards for visual display units in an insurance company

Operators of visual display units (VDUs) and microfilm reading machines at the Volksfürsorge insurance company are now protected by an agreement negotiated by their union in June 1979.

Under the agreement operators will be examined by a company-appointed optician before beginning such work and, after a follow-up eye test at the end of the first year, at regular 2-yearly intervals. Employees may also ask to see an optician if at all worried about deteriorating sight. Consultations and expenses will be paid by the company. If a transfer is considered necessary due to eyestrain, the operator must be offered suitable alternative employment. Expectant mothers will not be employed in this type of work.

As regards working conditions, the workplace layout will conform to accepted health and ergonomic standards, and any new developments will be taken into account. Not more than 6 1/2 hours a day should be spent on this type of work and there must be a 15-minute break after an hour's work. Wherever possible operators will also carry out other tasks which would effectively mean performing a "composite job". During the second quarter of 1980 full discussions will take place on the possibility of defining and establishing such "composite jobs" on a permanent basis.

The agreement also states that in all staffing and personnel matters, including choice of eye tests, the joint decision of the works council will be taken into consideration. (For other articles on VDU work stress, see S.L.B. 1/80, p. 90, 1/79, p. 84 and 3/78, p. 277).

Source: Agreement concerning the installation of video display units within the Volksfürsorge (Insurance Co-operative) Company (Hamburg), 21 June 1979, 4 pp.

ITALY

Olivetti agreement: restructuring and employment

The works agreement signed by Olivetti, the office equipment and industrial automation systems manufacturer, the unified federation CGIL-CISL-UIL and the Government follows the main lines laid down by the sectoral agreement concluded in July 1979 for metal trades as a whole (see S.L.B. 1/80, p. 43). It is also the first agreement to implement the provisions of Legislative Decree No. 624 of 11 December 1979 covering unemployment income maintenance and labour mobility in companies with economic difficulties (see S.L.B. 1/80, p. 68).

Because of financial difficulties Olivetti announced some time ago that it would have to make cutbacks affecting 4,500 employees, including 3,500 at its Ivrea factory.

The present agreement amounts to a regular restructuring and development plan which, with state and regional assistance and financing, should make it possible to avoid redundancies.

For the first time the unions have accepted the idea of early retirement. Any worker with a certain number of years' contribution may take early retirement five years before reaching pensionable age. The pension will be the same as that received at normal pensionable age, plus compensation, financed by the State, for loss of earnings. Two thousand wage earners have already asked for early retirement on these conditions.

Furthermore, 450 redundant workers are already entitled to benefit under the special system of allowances paid by the national wage compensation fund (Cassa integrazione quad agni - set up under Legislative Decree No. 624). During the whole lay-off period the workers will receive 80% of their wages and should be able to follow retraining courses organised by the undertaking with a view to redeployment.

The Government has promised to expedite implementation of the restructuring plan for the electronics and data processing sector and to finance a large-scale scientific and technical research programme in the same field. It has also said that it is prepared to accelerate installation of electronic and data processing equipment during the next 3 years throughout central and local government services, which will mean a considerable increase in state orders for such equipment. The orders will amount to some 60 million lire per year for the 3-year period 1980-82, including 40 million lire for a telex network (1 US dollar = 880 lire). It is intended to set up a centre to co-ordinate orders from the regional authorities as well as a higher institute of informatics. Funds will be made available for training skilled staff.

Olivetti intends to do everything possible to improve the competitiveness of its products nationally and internationally through a large-scale research and rationalisation plan. Among other things, it will engage 200 programmers to study the application of informatics to the requirements of public administration.

Four months after signature of the agreement it is already being contested by the National Metalworkers'

Federation (FLM), and a 2-hour strike was held at Olivetti-Ivrea on 2 April 1980. The FLM considers that the firm's position has improved enough for it to re-engage the 450 workers benefiting from the national wage compensation fund, as stipulated by the agreement; but the firm is challenging this obligation. The FLM also says that the State has not yet fulfilled its promises concerning the research and order programme.

The parties to the agreement should meet again on 30 October 1980 to review its implementation.

Source: Olivetti agreement (Rome), 21 Dec. 1979, 15 pp. (in Italian).
La Stampa (Turin), 25 Mar. and 3 Apr. 1980.

NEW ZEALAND

The arbitration court recognises the principle of prior consultation before the introduction of computer technology

The New Zealand Arbitration Court has ruled in favour of the Clerical Workers' Union in a case concerning the principle of prior consultation before the introduction of computer technology; this case is the first of its kind in the country. The Court ruled that the following clause be inserted in the collective agreement:

(a) where an employer is considering the introduction of new computer technology (including word processing machines) the employees likely to be affected by any such decision will be the first to be informed;
(b) where an employer has decided to introduce such technology the employer shall consult fully with the employees affected and the representative of the Union;
(c) when the introduction of such technology will result in redundancies, the employer shall notify the Union to enable discussions on redundancy to take place.

This is the first time that white-collar workers in New Zealand have had a technology clause inserted into their collective agreement.

Source: FIET: Newsletter (Geneva), No. 9, Sept. 1980, p.3.

Editor's note: The New Zealand Post Office Union has just brought out a study on the impact of new technologies in the postal and telecommunications services which will be analysed in the next issue.

NORWAY

New technology agreement reflects four years' experience

The Norwegian Federation of Trade Unions (LO) and the Norwegian Employers' Confederation (NAF) negotiated their first national collective agreement on computer-based systems in 1975. It was revised in 1978 to cover computer-based systems used for (1) the planning and execution of work and (2) the storage and processing of personal data.

While the agreement recognises that computer-based systems can be useful tools in the planned allocation of the total resources of an enterprise, it outlines practices to be followed in their introduction.

1. Computer-based systems must be evaluated, not only from technical and economic angles, but also from the social point of view, including impact on employment, work organisation and human relations.

2. Shop stewards and employees must be kept fully informed about the introduction of planned changes in such systems, so that they can express an opinion before any decisions are implemented.

3. All information must be provided clearly and in a language easily understandable to those without specialist knowledge of the subject concerned.

4. Furthermore, management and shop stewards, both separately and jointly, will provide employees with sufficient information for them to understand the fundamental features of the systems they use or which affect them.

5. Employees directly affected by the new systems should, to the greatest possible extent, be involved in the preliminary project work prior to introduction of the new system.

6. Employees may elect special representatives to safeguard their interests vis-à-vis the new technology. These representatives will have access to all information on the hardware and software being used. Training in general data-processing techniques will be provided by the employer and includes courses on systems analysis, programming and project administration sufficient to develop the competence needed for participating actively in system design. (Types of training course were added to the 1978 revised agreement.)

7. No collection, storage, processing and use of personal data will take place unless it can be objectively justified as being necessary for the work of the enterprise. The type of computerised personal data to be collected, stored, processed and used should be clearly specified.

8. Management, in co-operation with the shop stewards, should draw up detailed procedures for the storage and use of personal data. (The LO has drawn up model procedures.)

Local agreements may be drawn up within the framework of this general agreement and several have been signed, mostly in larger enterprises. One such agreement in the Telecommunication Company of Norway gives the unions the right to influence the actual purchasing of the computerised equipment. It also gives the unions the possibility to run their own programmes on the company's computer.

As a follow-up to the central agreement, the LO has appointed specialist and data-processing staff and, in co-operation with outside research staff, have published two basic textbooks for trade unionists.

Source: General agreement between the LO and NAF on computer-based systems, translation by the European Trade Union Institute (Brussels), May 1979, 6 pp.

Workers' participation in system development

Rapid technological change breeds new industries, creates new needs, demands new skills. Some of the adjustment problems this situation creates are already the subject of intense public debate as evidenced by the Second Conference on "Human Choice and Computers" organised by the International Federation for Information Processing (see S.L.B. 3/79, p. 237). While much of this debate focuses on the employment effects of micro-electronics and the invasion of privacy by too much information about individuals in data banks, one of the papers at the IFIP Conference explored the needs of trade unionists in dealing with the introduction of new technology.

Describing a research project carried out by the Norwegian Union of Iron and Metal Workers into the effects of computerisation, Kristen Nygaard, Research Director, Norwegian Computing Center and Professor of Informatics at the University of Oslo - author of the paper shows how traditional research provided no guidelines for trade union action. It was not enough to get a few researchers to find out something "on behalf" of the workers. A new strategy was therefore developed which redefined "research results" in terms of trade union actions.

The outcome was four reports, two of which were published as textbooks for the unions. Subsequently, there has been a central data agreement and several local agreements (see article above).

According to the author, the Norwegian experience highlights a number of points which ought to be valid elsewhere.

1. Trade union participation in developing computer-based systems implies a major extension or change in union policy. When the process starts it cannot be confined to what managers and/or system workers consider "useful", "justified" or "non-political".

2. Trade union participation cannot be built upon centralised activities within the unions or by relying upon hired "specialists". It must include its members and shop stewards.

3. Trade unions should not start their education in this area by acquiring the current knowledge and understanding of data processing of the system workers or the managers. Instead, they must start by building up their own basic understanding in terms related to their job situation and the trade unions' picture of the world. Additional factors can be taken into consideration afterwards.

4. Trade unions will experience a great number of defeats in this process of ongoing participation. They may, at other times, be irritatingly unreasonable in the opinions of managers and system workers. They may also experience serious problems within their own ranks and will have to start by emphasising a defensive strategy, before gradually becoming more constructive.

5. The extent and nature of trade union participation will also be dependent upon the major trends and events in industrialised market economies and production technology.

6. The situation of the system workers will continue to change drastically and they will get a less frustrating future if they start taking the word "participation" seriously.

In future, trade unions must broaden their interests even further to include the wider issues of job content, social contacts at work and power relationships.

Source: Norwegian Computing Center: "Workers' Participation in System Development", lecture given by Kristen Nygaard at the Second IFIP Conference on "Human Choice and Computers" (Vienna, 4-8 June 1979).

SWEDEN

Banks sign co-determination agreement

The first central agreement in the private sector under the 1976 Act on co-determination at work (see S.L.B. 2/76, p. 118, 3/76, p. 223, 4/77, p. 315 and ILO: Legislative Series, 1976 - Swe. 1) covers some 40,000 bank employees. Two agreements covering the public sector were signed in March 1978 and March 1979 (see S.L.B. 4/78, p. 342 and article above).

The agreement reached on 10 January 1979 by representatives of the Banking Insitutions Employers' Organisation (BAO) and the Swedish Bank Employees' Union (SBmf)

The agreement was reached on 10 January 1979 by representatives of the Banking Institutions Employers' Organisation (BAO) and the Swedish Bank Employees' Union (SBmf). Like the public sector agreement, it emphasises decentralisation of decision making and will be supplemented by local agreements.

Under the agreement the SBmf will be entitled to representation on the various decision-making bodies of the main bank. In addition, representatives responsible for co-ordination will be elected by the employees at individual branch or local offices.

The various areas for co-determination are listed in the agreement. Apart from manpower planning, work organisation and general conditions of work, they include: introduction or major alteration of data systems; the bank's business activities; mergers, transfers, closures and opening of banks; personnel policies and career planning; equal opportunity; policy guidelines for the distribution of work; the selection of candidates for different types of training and the establishment of job descriptions. Questions relating to the work environment and to safety and health are also covered.

Disputes are settled in the first instance through local or central negotiations, the final instance being a joint BAO-SBmf Confidence Board.

Source: BAO-SBmf: Collective agreement on co-determination in the banks, 22 pp. (in Swedish).

Newspaper industry: unions gain job security faced with new technology while guaranteeing "industrial peace" for 6 years

Following a new agreement in the Swedish newspaper industry on 1 May 1980, wage negotiations up to April 1986 will be carried out under an obligation to maintain "industrial peace". At the same time, a separate technology agreement guarantees that the introduction of new techniques will not lead to dismissals. Instead, journalists, graphical workers and clerical employees will be retrained during working hours to deal with the new techniques. A special education and training fund will be built up by the employers and retraining plans will be worked out by a joint council for the graphical industries.

Newspapers generally have to negotiate with trade unions representing respectively - journalists, graphical workers and clerical employees. The graphical union is affiliated to the Swedish Trade Union Confederation (LO) and the other two unions to the Central Organisation of Salaried Employees (TCO).

However, new computerised printing systems are threatening to break down the clear demarcation lines that exist between these three unions and under the new agreement existing jobs will change, becoming more integrated. For example, by typing 30% of the editorial text directly on the computer keyboard, journalists will be allowed to take over some of the traditional graphical functions. The intention is to improve quality and productivity and integrate existing skills into new types of jobs, not to have a one-way flow of traffic out of one union into another.

Under the agreement, these three categories of staff will continue to carry out their jobs in the same way as they do at present, but it will be possible for jobs to be transferred from one union to another after local negotiations. Local agreements must be approved by the national unions.

The current industry agreement was preceded by an inter-union agreement between the journalists and the graphical workers on 15 January 1980 by which the two unions agreed to respect each other's jurisdiction. They also agreed that new technology must not lead to redundancies. The co-operation between the two unions ensures a united stand in negotiating with the employers. While agreeing to co-operate with the employers to make newspapers competitive with other media, both unions are committed to the introduction of planning which will ensure job security, meaningful work and an improved working environment. They have also agreed to promote an industrial structure which serves the needs of the public for quality and free choice.

The working conditions of graphical workers are considerably better than those of industrial workers and the new agreement on technology is far ahead of anything obtained in the manufacturing industries. Under the peace agreement the graphical workers also won an extra vacation bonus from 1981 onwards - 3% of the total wage including bonuses and overtime.

Commenting on the agreement in its own union journal, the Graphical Workers' Union says the integration of jobs envisaged has already been introduced on the Norwegian newspaper "Ostlendingen", where the walls between the editorial and the composing rooms have already been broken down. This, it says, will hopefully in the long run lead to a one-media union.

Source: Grafiska Fackförbundet och Svenska Journalistf förbundet: Överenskommelse om ny teknik - Samarkete, 1980, 2 pp.
Fredsavtalet - Omröstning (Stockholm), 14 May 1980, 10 pp.

Social insurance workers accept computers but no deskilling

Sweden's social insurance workers - mainly women - have decided to face the problems of computerisation by increasing their own knowledge and skills and maintaining high quality social welfare services.

Faced with participation on a national commission of inquiry into how to handle computerisation in the social services, the insurance workers' union called in researchers and computer specialists from the Swedish Center for Working Life to help their representatives on the commission. The union has the right under the 1976 Co-determination at Work Act (see S.L.B. 3/76, p. 223) to do this for no cost.

The consultants and the union recommended a gigantic education programme, in the preparation of which 9,000 out of 18,000 members participated actively.

The programme that was subsequently drawn up starts by stating the moral basis for the union position as adopted by the Swedish Confederation of Trade Unions (LO).

It goes on to deal with certain aspects such as work organisation, employee participation in the development of the programmes and what is expected from a good system.

To maintain the quality of the service, the union demands that:

- skilled jobs in social insurance not be computerised;
- the computerisation of routine jobs be tried on a case-to-case basis to assess repercussions on service and working conditions;
- the professional knowledge of the personnel be maintained on a level that provides quality service and gives members opportunities for personal development through their work;
- training providing the staff with an overview of the various sectors of social insurance;
- more thorough research clarify the connection between computerisation and the deterioration of professional knowledge.

The union is also concerned about traditional safety and health issues, as well as demanding that the maximum working cycle for intensive visual display terminal work be limited to one hour - with a maximum of two such periods a day.

Source: Swedish Work Environment Association: "Working Environment 1981", Arbetsmiljö International (Stockholm), pp. 10-12.

SWITZERLAND

Electronics in newspaper offices: an agreement

In December 1979 the general meeting of the Swiss Federation of Journalists (FSJ) approved the joint agreement which its Central Committee concluded on 4 April with the Swiss Association of Newspaper Editors (ASEJ) and the Newspaper Union of French-speaking Switzerland (URJ) on the introduction and use of electronic equipment in newspapers and periodicals. This agreement is an annex to the current collective agreement and is subject to the same revision, conciliation and arbitration procedures. In view of the rapid evolution of technology, it has been signed for a period of two years only.

The main aim of the agreement is to protect the substance of the journalist's profession when electronic systems are introduced. Editorial staff must be informed and consulted at the start of any plans or work on installing such systems. The journalist himself decides whether it is technically advisable to use electronic news-gathering equipment to do his job, which cannot be limited to working on an electronic console. The system must be designed to allow journalists to make any necessary text changes before publication, and to obtain a printout of the copy fed into the computer memory.

The agreement is also designed to protect the health of journalists, and provides for compulsory eye examinations in the early weeks of working on visual display units and thereafter every year.

Finally, the agreement is designed to protect employment in the profession. An editor can only terminate a journalist's contract because of the introduction of electronics if it is impossible to propose an acceptable equivalent job. If made redundant, the journalist is entitled to one year's "mobility compensation" equalling the difference between the previous salary and what is earned in the new job or between the final salary and unemployment benefit.

Any journalist required to use electronic equipment must be given proper training which will count as work time.

Continuous training was the subject of an agreement for French-speaking Switzerland signed on 8 June 1978 by the FSJ and the URJ which set up a joint commission responsible for organising continuous training for journalists and managing a fund set up for the purpose. The fund is financed by equal contributions from journalists and editors (1% of the base salary). A similar agreement is being considered for German-speaking Switzerland.

Source: Accord paritaire ASEJ/URJ-FSG sur l'introduction et l'application de l'électronique dans les rédactions, signed at Berne, Lausanne and Zurich on 4 Apr. 1979, 4 pp.
Accord paritaire URJ/FSJ sur la formation continue des journalistes professionnels, signed at Lausanne on 6 June 1978, 7 pp.

UNITED KINGDOM

Ford reaches agreement on computerisation

On 13 September 1978, the Ford Company and the unions representing its workforce reached an agreement on the introduction of computers which contained the following provisions:

1. No dismissals as a result of computerisation.
2. Salary levels to reflect increased productivity derived from the introduction of the computer.
3. Shift work resulting from computerisation not to be required of staff in jobs where it is not currently accepted unless agreement is reached with the union concerned or until all negotiating procedures have been exhausted.
4. Additional skills required to use computer equipment to be reflected in salary grades.
5. The provision of adequate training for personnel called upon to use computers and for the redeployment of those displaced by computerisation.
6. Maintenance of the company commitment not to ask employees on hourly wage scales to carry out salaried work. Any change in this policy concerning computer work is subject to consultation with the unions.
7. Retrenchment in occupational groups due to the introduction of computers to be subject to prior consultation with the unions.
8. Discussions to be held between the Company's chief medical officer and the medical officer of the Trades Union Congress, on the health implications of using visual display terminals, etc.
9. Information acquired specifically or incidentally by computer-based systems shall not be used for individual or collective work performance measurement.
10. Prior negotiation with the union or unions concerned on any change or extension of the use of computers which affects employees.

Within the framework of this general policy, each union will be responsible for negotiating appropriate arrangements in respect of the occupations it represents.

Source: The TGWU Record (London), Nov. 1978.

Post Office workers negotiate security of employment

Over the next decade at least, current manpower totals in the Post Office can be maintained in face of rapid modernisation. This is the message of an exhaustive study by the Research Department of the Post Office

Engineering Union (POEU), of all the technological developments that are likely to affect Post Office telecommunications in the next ten years.

The POEU has reinforced this conclusion by negotiating with the Post Office a security of employment agreement which guarantees that there will be no compulsory redundancies as a result of modernisation and the agreement, with the exception of one clause (No. 5), was accepted by the POEU Annual Conference on 3 to 8 June 1979. The clause which failed to get Conference approval qualified the job security provision by making it ineffective in the event of manpower problems arising due to causes outside the control of the Post Office. This clause is being renegotiated.

The agreement outlines practices for the planned introduction and exploitation of new technology which consist of: (a) frequent updating of manpower forecasts; (b) adjustment of manpower supply through natural wastage, the regulation of recruitment, reduced overtime and staff redeployment; (c) extending work of POEU members to that usually carried out by contract labour.

A special redeployment unit for surplus staff has been set up under the agreement which also sets out mechanisms for anticipating local manpower surpluses well in advance and for planning for the consequences of such surpluses through detailed procedures for retraining, redeployment and relocation.

All measures for dealing with manpower surpluses will be integrated into the normal manpower planning process and subject to continuous review. Apart from cutting back on recruitment and overtime, they include temporary or permanent transfers, reassignment to lower grade vacancies, and voluntary retirement, accompanied by redundancy payments.

When deciding the suitability of alternative employment for staff, there should be full consultation with the POEU branches concerned, and individual circumstances should be taken into account, particularly for the disabled. Where alternative employment regarded as suitable is offered to staff, they should be advised that where this is unreasonably refused, they will not be eligible for compensation within the terms of either the Post Office or state redundancy schemes.

After almost a year of industrial action in 1977/78, the union secured a 37 1/2-hour week from 1 December 1978 and this is being taken in a variety of ways which includes a 9-day fortnight for a minority. The POEU will continue to link a shorter working week to technological change and the claim for a 35-hour week remains on the negotiating table.

Source: POEU: The modernisation of telecommunications (London), June 1979, 104 pp.
Job Security Agreement with POEU, 9 pp.

Technology agreement signed at Northern Engineering Industries

A collective agreement has been concluded between APEX and a branch of Northern Engineering Industries on the introduction of visual display units (VDUs) and associated systems such as desk-top data entry devices and other computer linked equipment including word processors. The agreement provides for: (1) job security - no redundancies; (2) a 5-day training course covering a jointly-agreed syllabus on VDUs and their health and safety implications, with a special payment to those who attend the course; (3) 6-monthly eye-tests for VDU operators; (4) a 20-minute break after one hour's work at a VDU; (5) a joint management/union team to monitor the further development of computer-aided systems.

Source: Industrial Relations Review and Report (London), No. 202, June 1979, pp. 15-16.

ITV settlement guarantees consultation before introducing new technology

Independent Television (ITV) companies went back on the air in the last week of October 1979, after a blackout lasting 11 weeks.

The emphasis of discussions changed markedly during the dispute. Starting as a straightforward pay dispute, it increasingly focused on the introduction of new technology, the unions, particularly the Association of Cinematograph, Television and Allied Technicians (ACTT), seeing the use of electronic news gathering (ENG) equipment as a threat to jobs.

In the final over-all agreement, the unions won a salary increase amounting to 45% by summer 1980 plus four extra days' holiday over a two-year period. However, they also agreed to assist the introduction of ENG techniques through local negotiations with a view to concluding technology settlements by next April. If negotiations fail, local-level disagreements on introducing ENG will go to a tribunal.

These technology agreements will be governed by a code of practice agreed between ACTT and ITV. Under the code, prior consultation must include consideration of manning levels, grades of the employees involved, training and retraining.

The aim should be to achieve an agreement either for the introduction of the new technology or for a trial period.

Source: ITV/ACTT Code of Practice on the introduction of automated and new equipment, 3 pp. (effective 1

July 1979).
Financial Times (London), 18 Sep., 20 Oct. 1979.
The Economist (London), 27 Oct. 1979, pp. 27-28.

Technology agreement signed at Vickers

A technology agreement covering the introduction of visual display units (VDUs) and remote batch terminals (RBTs) has been concluded between the Association of Professional, Executive, Clerical and Computer Staff (APEX) and part of the Vickers group.

While not going as far as a previous APEX negotiated agreement (see S.L.B. 3/79, p. 233) the settlement provides for:

(1) job security (no redundancies due to retrenchment);
(2) the introduction, on a continuing basis, of improvements to safeguard the health and safety of the operator as and when they are developed;
(3) agreed lighting and flicker standards;
(4) eye tests, at the company's expense, upon request from the operator;
(5) comfortable positioning of keyboards.

The agreement also specifies pay rates for VDU and RBT operators - an area in which precedents are still being set. Rates for VDU operators were increased to 3,400 pounds, from 2,760 pounds on 1 June 1976. Rates for RBT operators went up to 3,500 pounds, from 2,864 pounds (1 US dollar = 0.45 pounds sterling).

Source: Industrial Relations Review and Report (London), No. 207, Sep. 1979, pp. 13-14.

Times newspaper dispute ends after 11-month suspension: technology key issue

The management of the Times Newspapers Limited reached an agreement on Sunday, 21 October 1979 with the National Graphical Association (NGA), making it the last union to settle in a complex dispute that kept the paper off the bookstands 11 months. The agreement came only hours before a management imposed deadline, after which all print workers were to be dismissed and the paper discontinued.

Publication of The Times, the Sunday Times and three monthly supplements was suspended on 30 November 1978 by the management after it failed to reach agreement with the unions on reduced manning levels (though there would be no compulsory redundancies), a disputes procedure to eliminate wildcat strikes and general acceptance of new technology. The print unions were adamant in refusing to operate computer-linked type-setting equipment and, in particular, to allow journalists and advertising staff to use it.

Negotiations were complicated by the fact that management not only had to negotiate with the unions at national level but also with 54 internal semi-autonomous union branches directly representing the staff of 4,300.

Direct intervention by the Employment Secretary to initiate a settlement in December 1978 and in March 1979 failed to produce a peace formula. In April 1979, an attempt to produce a European weekly edition of The Times in Frankfurt (Federal Republic of Germany), using Turkish compositors, was abandoned due to the solidarity of the German print unions with their British counterparts.

In June 1979, the NGA - one of the main print unions - decided to find new jobs for its 600 members. This sparked off a personal meeting between the chairman of the International Thomson Organisation, which owns Times Newspapers and the general secretary of the NGA. After a four-hour meeting it was agreed that technology was no longer an obstacle. Management agreed to resume publication before getting agreement with the NGA on the company's demand that journalists and advertising staff, as well as NGA members, must eventually be allowed to operate the computerised composing system.

The NGA, for its part, agreed to enter into talks on new technology with the management and with the two other unions involved - the National Union of Journalists (NUJ) and the National Society of Operative Printers, Graphical and Media Personnel (NATSOPA). The intention was to reach a "conclusion satisfactory to all parties" within 12 months.

In August, talks between the newspapers' management and trade union leaders collapsed again as the new conditions were rejected by several NATSOPA bargaining units representing a total of 2,500 workers. By October, NATSOPA had agreed to the settlement and negotiated its own new pay agreement.

This new pay agreement eroded the skill differentials (higher payment for higher skills) vis-à-vis NGA members and, on 18 October, negotiations with NGA broke down over the restoration of these payments.

Concessions were made on both sides in the last-minute 26 hours of non-stop negotiations and the final results are outlined below:

1. New technology. Agreements to operate new composing equipment have been signed. The equipment already installed is designed for direct access by journalists and clerical staff, but agreement on this has not been obtained. That issue will be the subject of joint discussions over the next 12 months. A single composing room will be used for all five publications, which will move to computer setting over the next two years.

2. New dispute procedures have been agreed but these have no penalties attached to them. They set out five levels of appeal during which no industrial action or lock-out will be permitted. Guarantees of continuous production during the appeals procedure have been signed.

3. Manning levels. Agreements foresee the gradual loss of 600 jobs. Staff cuts range from up to 44% in the composing room, through 27% in publishing to 20% for NATSOPA machine assistants. Negotiators were unable to reach agreement on NGA machine managers. This issue has gone to binding arbitration by the Advisory, Conciliation and Arbitration Service (ACAS). Management is seeking a cut of at least 100 clerical staff from the 560 not yet covered by the agreement. Shifts are being reduced by 30 to justify a pay rise.

4. Pay rises average 40-45% immediately with a further 5% in October 1980 - which is in line with similar rises elsewhere within the industry. Piece rates for linotype operators will eventually be replaced by time rates, although the terms have yet to be negotiated. There have also been improvements in overtime pay; introduction of a standard 6-week annual holiday and compositors will move gradually towards a 32-hour week. In addition, 7 million pounds (1 US dollar = 0.45 pounds sterling) has been set aside for voluntary redundancies.

In all, the shutdown cost 30 million pounds.

Source: Financial Times (London), 30 June, 6 July and 8, 13, 22, 23 and 24 Oct. 1979.
The Economist (London), 20 and 27 Oct. 1979.

Technology agreement at Plessey Telecommunications

A technology agreement concluded between Plessey Telecommunications and the Association of Professional, Executive, Clerical and Computer Staff (APEX) takes effect as from 11 July 1979, covering 4 plants and some 600 employees.

Under the agreement, equipment has been introduced to control the manufacturing process and provide accounting information. It includes 31 visual display units (VDUs). The main provisions include: (1) the provision of sufficient information to enable the union to monitor the effects on jobs, health and safety, (2) no redundancies as a result of the new systems, (3) training for APEX members in the user departments, (4) no unilateral extension of shift working, (5) 6-monthly eye tests for VDU users, (6) maximum standards for VDU glare and regular rest breaks.

The parties also agreed that the maximum flexibility shall exist to develop working methods that provide satisfying jobs and that an agreed positive programme of manpower and job development will be introduced with the object of increasing the responsibility and autonomy of job groups and employees. (For similar agreements see S.L.B. 3/79, pp. 229 and 233 and 4/79, pp. 328-330.)

Source: Industrial Relations Review and Report (London), No. 215, Jan. 1980, pp. 15-16 (includes text of agreement).

New ASTMS technology agreements in insurance and aeroengineering companies

A new technology agreement between the Royal London Mutual Insurance Society and the Association of Scientific, Technical and Managerial Staffs (ASTMS), provides a framework for joint consultations on technological change.

Under the agreement the Society agrees not to introduce new technology or new techniques at a rate which may cause compulsory redundancies. Any staff member whose job becomes redundant will be offered suitable alternative employment. It also provides for built-in union representation at various stages when changes are planned, as follows:

- where possible, two union representatives to sit on managerial steering committees evaluating the possible application of new technology;
- the setting up of a permanent joint body (a "technology conference") having equal management-union representation to discuss changes recommended by the steering committees;
- the referral of all matters which are subject to collective bargaining to the joint negotiating committee;
- status quo until agreement has been reached or procedures exhausted;
- union access to detailed information on manpower and skills requirements, necessary training, and safety and health problems;
- where possible, time off for ASTMS representatives on the Technology Conference to attend trade union or trade union approved courses.

A similar agreement with Rolls-Royce aeroengine company supplements existing negotiating machinery by setting up a small central committee, to meet quarterly for joint discussions on the total spectrum of system changes. The company will supply union representatives with information on possible changes as far in advance as possible. The Committee will try to identify problems for employees which may arise from equipment/systems changes. Joint discussions will then take place at the plant in which a systems change is to be introduced. Representatives from the

Central Committee can be co-opted to take part in these discussions.

The status quo will be observed during any negotiations. Both sides agreed some training in systems may be necessary (in-house, external or trade union-sponsored). Job security was dealt with under a separate manpower agreement.

Both agreements establish very broad definitions of new technology. They encompass any form of technological or systems change rather than concentrating on one piece of equipment. Technology is moving so fast that limited agreements would rapidly become out of date.

Source: Agreement between the Royal London Mutual Insurance Society Limited and ASTMS, Feb. 1980, 4 pp.
Agreement between Rolls Royce Limited, Derby Group of Factories and ASTMS, 2 Feb. 1980, 2 pp.
Industrial Relations Review and Report (London), No. 219, Mar. 1980, pp. 9-10.

APEX agreement gives on-going consultation on new technology

An agreement covering the introduction of new technology has been signed between the International Harvester Company and the Association of Professional, Executive, Clerical and Computer Staff (APEX).

It goes further than some of the agreements already signed (see S.L.B. 3/79, p. 233 and 4/79, p. 329) in that it is backed up by a separate operating agreement on visual display units (see below p. ..), provides for joint consultataion on the long-term manpower implication of new technological equipment, and a guarantee of no compulsory redundancies. In cases of re-deployment to a lower grade, employees will keep the same grade.

The agreement also specifies workplace environmental standards and the signatories agree that any precise definition of new technological equipment may be an unreasonable restriction on both sides.

Source: Agreement between International Harvester Company of Great Britain Limited and APEX, 4 Feb. 1980, 2 pp.

CBI decides not to ratify the technology pact with TUC

The draft agreement on new technologies which was approved by representatives of the Trade Union Congress (TUC) and the Confederation of British Industry (CBI) in

July and endorsed by the TUC Congress in Brighton in September, was rejected by the CBI Board on 15 October 1980.

For several years the TUC has been inviting its members to conclude this type of agreement with employers so as to set up collective bargaining machinery to deal with problems of job security and working conditions brought about by technological development and to reach solutions on a joint basis (see S.L.B. 3/79, p. 230). The result has been that several companies have already reached such agreements (see, for example, S.L.B. 4/79, pp. 329-330 and 1/80, p. 8) and the TUC was now hoping to conclude a nation-wide agreement covering the whole of British industry.

The technology agreement which had been envisaged by the representatives of the TUC and the CBI is in fact a joint declaration of the inevitable, necessary and positive nature of technological change for British industry. It also recognises the risks which such change poses to the level of employment, the skill content of jobs and working conditions. It concludes that there is a need for employers, workers and their trade unions to participate jointly in the choice of new technology, in its application and in the distribution of any benefits.

This participation should cover full employee access to information and, where applicable, the appropriate training of employee representatives, as well as regular consultation on the effects of new technologies by setting up some sort of joint machinery within the company.

Employers should make every effort to guarantee job security where new technologies threaten manning levels and, as part of a policy of systematic manpower planning, to give employees adequate opportunity for acquiring any necessary skills through training or retraining.

Productivity gains resulting from the adoption of new technology should improve the conditions of work of manual and non-manual employees (hours, annual leave, pensions) and lead to a reorganisation of general working patterns (for example, less shift work and overtime).

Lastly, employees should be involved in monitoring the hazards that new technologies post to their health and safety, especially during the trial and initiation periods following the introduction of new production methods. Attention should also be given to modifying the established grievance and disputes procedure accordingly.

Many CBI members had supported this agreement on the basis that it simply confirmed established practice and had the advantage of allaying union anxiety and reticence about technological changes which were inevitable if the competitive position of British industry was to be assured.

However, the majority opposed the agreement (especially in the manufacturing industry) and saw a risk of the guidelines being used by the unions to block even minor innovation. They should therefore be reserved for major technological decisions.

This rejection is likely to chill relations between the two federations.

Source: Trade Union Congress: Technological change, TUC-CBI statement. TUC Publications (London), Aug. 1980, 7 pp.
Financial Times (London), 23 July, 13 and 16 Oct. 1980.

New technology: review of APEX agreements

The Association of Professional, Executive, Clerical and Computer Staff (APEX) has been in the vanguard of negotiating technology agreements with employers in a range of industries. A recent review published by APEX sums up the general framework of these agreements. Some define technology very broadly, refer specifically to visual display units (VDUs) or, in a few cases, to named systems. The scenario varies. Some agreements are tied in with wages and productivity, others are completely separate.

Agreement prior to introduction. Some agreements include an explicit guarantee of no introduction or extension of systems or units without prior union agreement or negotiation. In other cases, only consultation is assured. The majority incorporate a reference to existing procedures for settling disputes.

Monitoring. All agreements include a management guarantee to provide detailed information. It is not clear in many cases, however, when and how this is to be disclosed or how the APEX representatives propose to assess this information and monitor developments. While one company has set up a special management/APEX committee, many companies use existing machinery.

Job security. No agreement to date incorporates a "no job loss" clause. Some guarantee no redundancies as a direct result of the introduction of new systems, coupled with a commitment to negotiate redeployment where a labour surplus arises. Other agreements only guarantee no compulsory redundancies and, in some cases, specifically allow voluntary redundancy, early retirement or natural wastage. A few refer the matter to existing procedures.

Improvements in hours and conditions. Most of the agreements contain no reference to reduced hours or improvements in other conditions. The most positive statement is in the Express Lift Company agreement where: "The management agrees to negotiate with the union on the union's target of ... a reduction in the working week and improvements in holiday arrangements in the light of increased productivity as the result of the use of the systems".

In three other companies, including Rolls Royce,

management formally recognised the union's right to pursue improvements in the conditions of employment to take account of increased productivity arising from new systems.

Job content. The majority of agreements include a guarantee that no employee will be downgraded and referral of revised jobs to joint Job Evaluation (JE) machinery. A few go further and include the possibility of review or renegotiation of the JE scheme if this is found to be necessary. No agreement contains guarantees that jobs will not be downgraded. At International Harvester and Rolls Royce (Crewe), redeployment to a lower grade job is permitted but with the individual guaranteed the same salary grade and increments. Some include clauses aimed at minimising routine and monotony in revised jobs. One agreement is unique in including grades, payment and duties of the VDU operators.

A common provision is an assurance that staff will not be required to undertake extra tasks or responsibilities without prior union consultation and/or agreement. Some, however, accept management-defined "flexibility".

In over half the agreements, no work would be subcontracted as a result of new systems without prior union agreement. All new shift-working proposals are also generally subject to union consultation.

The question of whether work performance of individuals is to be measured by computer-based systems is not dealt with consistently. Some companies have given formal guarantees not to use such systems while, in others, automatic monitoring of performance has been accepted.

Training. The most comprehensive experience of training in new systems among APEX representatives is probably at Northern Engineering Industries (see S.L.B. 3/79, p. 233) whose initial training scheme is attached to the substantive agreement. Most other agreements include more or less detailed references to training.

Health and safety. These issues take up the greater part of many of these agreements and are generally spelt out in far more detail than other important aspects of employment.

Many employers state that it is their intention that employees do not spend extended periods at VDUs and users are often divided into casual/occasional, regular and/or heavy/continual. Rest pauses are usually allowed only for continual users and provisions vary as follows: 60 minutes away from the screen each day, to be taken at regular intervals; a minimum of 10 minutes after every 60; 10 minutes in 50; 15 minutes after 60; 20 minutes after 60; an unspecific break after 40 minutes.

Almost all these agreements include a clause recognising that health and safety and ergonomic standards can be improved over time, so enabling renegotiation on these aspects, as well as on minimum lighting, noise, screen requirements, etc.

Source: APEX Research Department (mimeo.), 1980, 9 pp.

Technology agreement in television manning levels maintained

A major stumbling block in the 1979 strike by some 5,000 television technicians (see S.L.B. 4/79, p. 328) was the introduction of electronic news gathering equipment (ENG), which can be operated by a basic team of two people. Its advantage is simply a question of miniaturised micro-technology replacing rather cumbersome, conventional film cameras. It is particularly suitable for news coverage but its introduction does not rule out the use of conventional camera crews for other types of TV reporting.

The agreement signed by the Association of Cinematograph, Television and Allied Technicians (ACTT) with Independent Television after six months of negotiation outlines the manning levels of both operators and associated editors, plus changes in salary structure. They have obtained a commitment that not only will there be no redundancies but that present manning levels will be maintained. This means that any technician who leaves will be replaced. If, at a subsequent date any crew members are not necessary, any reduction of employees is to be based on natural wastage.

All appointments are to be made from existing film technicians, recordists and editors. Existing staff and future recruits will be trained to use both ENG and conventional film equipment. A small sub-committee with two representatives from both sides of industry will discuss initial training plans.

This training will enable employees to use both techniques and facilitate redeployment and flexibility between the various types of programme.

The company has agreed to give staff crews "a fair and reasonable share of film coverage", so protecting the amount of work given to in-house staff vis-à-vis freelance work. Statistics on ENG assignments will be produced quarterly to monitor its introduction.

Source: Industrial Relations Review and Report (London), No. 233, Oct. 1980, pp. 7-10.

Model technology agreement for banks and insurance companies

A booklet on microelectronics, which includes guidelines on the kinds of agreement banks and insurance union officials should be seeking on new technology, has been published by the Banking, Insurance and Finance Union

(BIFU).

The 32-page booklet gives details of a model new technology agreement the union would like to sign with all the companies it deals with, guidelines on the use and specification of visual display units (VDU) and a sample "assessment form".

The forms would be used to assist union members in monitoring developments at the workplace, to encourage their involvement in this process and to keep national union officials informed about what is happening. It contains sections on "reasons given" for the introduction of new equipment, health and safety implications and repercussions in terms of job content, job losses, redeployment, skill changes, retraining and wages.

The union is worried about the implications of new technology, particularly any possible repercussions on jobs, and recommends negotiating reduced working hours to ensure work sharing. This would be done by working towards a 4-day 28-hour week, a minimum of 5 weeks' holiday, and voluntary early retirement. BIFU also wants to see more sabbatical and educational leave.

The booklet, which lays down the broad strategy, is partly the product of annual conference decisions and the work of the union's microtechnology committee.

Source: BIFU: Microtechnology - a programme for action (London), Jan. 1981, 32 pp.

UNITED STATES

Technology vs. jobs: newspaper, postal and railway settlements

Principal issues in recent industrial strikes in the USA have revolved around job security in face of technological changes.

Newspaper settlements. The "New York Times" and the "Daily News" resumed publication on 6 November 1978 after a strike lasting 88 days. The "New York Post" had already resumed publication on 5 October after promising to abide by the settlement of the two other papers. Print men struck after the publishers decided to reduce manning levels by 50%, claiming that technological improvements had eliminated the need for some jobs. The strikers were supported by other unions representing editorial and clerical employees. The final agreement, although including wage settlements, essentially provides that the jobs of the 1,500 regular printers at all three newspapers are guaranteed until March 1984, while the publishers are entitled to reduce manning levels through attrition and to reduce overtime work.

Railways. In January 1979 the Brotherhood of Railway and Airline Clerks (BRAC) signed an agreement with the Norfolk and Western Railway, ending a dispute over automation and job security dating back to 1976. Under the settlement, reached after an 82-day strike, those with 3 years' service are protected against loss of wages caused by abolition of jobs. This agreement is separate from negotiations between the carriers and all the major unions (including BRAC) on reducing the size of train crews. The vital issue in the negotiations was the decision by management to eliminate one out of two brakemen on diesel trains - saying that these manning levels, which had been agreed before introduction of diesel engines, were no longer necessary. In a package deal, including improvements in medical benefits and holidays, most major unions have agreed that carriers will be allowed to eliminate one brakeman from trains with fewer than 70 cars, but that the rundown in the workforce must be achieved by normal attrition. At one point, the dispute completely shut down the railroads for four days.

Postal services. Members of the largest postal union overwhelmingly ratified a new contract with the Postal Service in October 1978, thus ending a long-drawn-out dispute involving over half a million workers. The dispute, finally settled by mediation, included a clause protecting existing workers against redundancies from technological change - a major union bargaining priority. However, the Postal Service may lay off new employees until they have 6 years' accumulated service and mandatory overtime provisions were also retained.

Source: The Washington Star, 23 Aug. 1978.
The New York Times, 7 Nov. 1978.
BNA: Collective Bargaining Negotiations and Contracts (Washington, D.C.), 6 Apr., 27 July, 21 Sep., 19 Oct. and 16 Nov. 1978.
AFL-CIO News (Washington, D.C.), 13 Jan. 1979.
US News and World Report (Washington, D.C.), 9 Oct. 1978.

Stock ownership, better pensions, time off and technology feature in General Motors-UAW contract

Under the three-year contract signed between the United Auto Workers Union (UAW) and the General Motors Corporation in September 1979, 450,000 employees will get bigger wages and cost-of-living allowances, and a 40% increase in pension benefits, more time off for training for new technology. As with other contracts negotiated by the UAW (see preceding article and p. ..), inflation protection for pensioners was a high priority for negotiators. The contract is based on a projected inflation rate of 8% per year. For the first time there will be automatic pension increases in each year of the agreement. And with increased retirement benefits, many now working may decide to retire early.

Another breakthrough marks an important step towards eliminating the double standard between manual and white-collar workers. A free stock ownership plan has been introduced which is identical to the one now available to GM salaried workers. Dividends will be credited to individual accounts, to be reinvested in additional shares - which are held in trust until retirement or termination.

Paid holidays are increased by an extra 26 personal holidays with pay spread over 3 years (up from 12 over 2 years). This is another step towards the UAW's goal of an eventual four-day working week.

New technology was dealt with in the agreement by the introduction of a new procedure aimed at reducing the impact of job erosion. It includes adequate advance notice to the union of any anticipated technology changes and the introduction of training programmes for skilled workers covering programming, computers, and numerically controlled machines and equipment.

Other benefits covered occupational safety and health (ranging from improved benefits to greater protection at the workplace), improved seniority rights, better disability insurance, and survivors' benefits, and further improvements to strengthen unemployment funds in the face of mounting layoffs in the industry.

Special provisions are included to cover newly recruited workers, who will be recruited at 60 cents an hour less than the base rate for the first month and at 35 cents less for the next two months. In addition, certain sickness and accident benefits will not be granted until after four months' employment, and full benefits after one year.

As regards absenteeism, joint discussions will be carried out to consider pilot projects for evaluating causes, effects and remedies, and to examine alternative work schedules.

Source: UAW: Report on the UAW-General Motors 1979 Tentative Settlement (Detroit), 18 Sept. 1979.
The New York Times, 23 Sept. 1979.

Script-writers win share in profits from new technology

Following a protracted 13-week strike, the Writers' Guild of America, which represents film and TV script-writers, concluded a new collective agreement that takes into account the fact that, due to new technology, scripts can be used many times.

In the past the use made of a script was more or less measurable. Incorporated in a film or tele-film, it would be used a certain number of times - for the original film, for television broadcasts and for re-runs. Estimates on the number of times it would be used were the basis for script-writers' fees. With the development of cable television and of video-cassettes, the use made of a script became difficult, if not impossible, to measure. Script-writers were faced with the problem of how they could be remunerated in accordance with the number of times their scripts were used. Other artists, film and television actors, musicians and film makers similarly affected have also resorted to strikes in order to reach agreements on this question (see S.L.B. 4/80, p. 422).

The agreement negotiated by the Writers' Guild of America is based on the same system of payment as that agreed to by the actors. It provides for the payment of a certain percentage of the receipts earned by the producers, once production costs have been covered. The difference between the script-writers agreement and the others, however, is that the script-writers have obtained a fixed percentage (2%) that is based on a combination of receipts from both cable television (or pay TV) and video-cassettes. The other agreements had separate formulas worked out for cable TV and for video-cassettes. And unlike the other agreements, there are no reserve clauses or references to the number of times a film is used. The actors, for example, are to be paid a royalty only after a film has already been shown for 10 days during the first year. It has been estimated that few films will be shown more than that number.

These agreements are relatively novel in the area of the arts and entertainment and illustrate the growing problem of providing for equitable fees in the face of new technology that enables works to be endlessly re-used.

Source: International Herald Tribune (Zurich), 13 July 1981.
Le Monde (Paris), 19-20 July 1981.

NORDIC COUNTRIES

Bank employees demand co-determination to deal with technological change

The rapid development of technology in the banking sector is a cause of concern to bank employees in the Nordic countries - primarily because of the implications for deskilling levels, but also because of uncertainty about changes in working conditions and work assignments.

For these reasons the Nordic Bank Employees' Union (NBU), with a total of 128,000 members throughout Denmark, Finland, Iceland, Norway and Sweden, is demanding that employees be associated with decisions taken in the planning stage whenever new technical systems are introduced. Such co-determination should figure in an agreement that ensures workers' representation in all decision-making bodies and working groups.

The NBU action programme contains demands for a successive reduction of weekly working hours, first to 35 and eventually to 30 hours. The programme also calls for better protection of bank employees against robberies and injury, increased job satisfaction and any necessary retraining.

Another point being considered by the NBU is an eventual demand for a right of veto, should the union feel that the interests of the staff have not been properly met.

Source: Nordiska Bankmannaunionen - NBU (Stockholm), 20 Mar. 1980.

EUROPEAN COMMUNITIES

Tripartite discussions take up the question of microelectronics

A major report of the European Commission, discussed by the European Council on 29 November 1979 called for a Community approach to the challenge of new technology. This was necessary not only to maintain the competitive position of EC countries but also to develop a coherent social policy in view of the inevitable consequences on employment, working conditions and industrial relations.

Such an approach would include: closer co-ordination of national research and development programmes especially in the space, telecommunications and information processing field; the establishment of common technical standards; public procurement policies which offer EC countries a degree of preference over outside competitors; harmonisation of data privacy legislation; EEC support programmes for national projects; public information campaigns and training programmes for specific groups.

Since then a follow-up report by the Commission on the impact of microelectronics was discussed on 26 February 1980 by the EC Tripartite Standing Committee on Employment (comprising national Labour Ministers, representatives from both sides of industry and Commission officials).

After studying the European Commission's latest thinking on the subject, the Employment Committee does not believe that computerisation will necessarily lead to a net loss of jobs but neither is it sure that it will create sufficient new jobs to compensate fully for those that will be lost. However, it was essential, when the programmes of principal world competitors are receiving massive public aid, to ensure that European industry as a whole has access to new technologies in good time.

The Committee therefore saw a need to develop Community-wide responses to the problems created by new technology. Among other things these responses must include:

- a continuing dialogue between management and workers both at Community and national levels;
- an increase in the rate of economic growth in the Community, measures to counteract inflation and the establishment of reorganisation policies for regions and economic sectors especially affected by technological change;
- technology-related improvements in the quality of life especially for disadvantaged groups such as the handicapped;
- stronger social protection measures to offset any negative effects on the standard of living of workers, particularly those displaced by technological developments and having difficulty in obtaining alternative employment;
- direct information campaigns geared towards the general public and also to specific interest groups.

Training was seen as the key to society's adaptability to future change. There was a need to change the basis of general education and to develop training adapted to the new technology.

While the workers' representatives (European Trades Union Congress - ETUC) emphasised the need for full trade union involvement in the introduction of technology, the employers (Union of Industry of the European Community - UNICE) saw effective adaptation to technical progress as being dependent on respecting the independent role of industry. Faced with the threat of job losses due to technology, the workers' representatives stressed the need to reduce working hours as a means of employment creation. The Committee felt that any follow-up to the Council Resolution of 18 December 1979 concerning the adaptation of working time (see S.L.B. 1/80, p. 86) should take into account the possible employment effects of new technology.

As a result of the discussions, a programme of activities within the EC was agreed consisting of: (1) Research into the qualitative and quantitative effects of microelectronics on employment, including an "observation mechanism", involving trade unions and employers, to monitor developments and ensure rapid response to urgent needs. (2) Setting up a "Community Research Pool" in order to centralise data on the impact of new technologies. (3) The adoption of employment policies encouraging geographical and occupational mobility among workers. (4) Further consultation at European level on the basis of short and medium-term employment forecasts.

Source: European Communities: Standing Committee on Employment: Press Release No. 5274/80 (Brussels) 26 Feb. 1980, 15 pp.
Commission of the European Communities: (1) European Society faced with the challenge of new information technologies: a community response,

Doc. COM(79) 650 final (Brussels), 26 Nov. 1979, 50 pp.; (2) Employment and the new microelectronic technology, Doc. COM(80) 16 final (Brussels), 5 Feb. 1980, 27 pp.
FIET: Newsletter (Geneva), No. 4, Apr. 1980.
Comité de Liaisons d'Employeurs: Communiqué de Presse (Brussels), 26 Feb. 1980, 2 pp.

EUROPE

Collective bargaining in Western Europe: a systematic survey of trade union action

The first attempt at a systematic survey of collective bargaining developments in 17 Western European countries was published in June 1980 by the European Trade Union Institute (ETUI). Based on replies to national questionnaires, it shows how the trade unions have reacted, against a background of economic recession, in defending the interests of their members, and how the scope of collective bargaining has changed.

While the different situation in each country is shown in the country analyses, the report shows that most trade unions will continue to focus on the following general priorities in collective negotiations: job security; job creation; a reduction of working hours; preservation of workers' spending power; the improvement of social protection for workers; improved working conditions, with special efforts to promote workers earning low wages.

The democratisation of the economy will also be an important issue.

As regards general policy, the European trade union movement considers that a new perspective is necessary: "What is needed in order to cope with the crisis is an active employment policy, structural policies (at both regional and sectoral level), reform of the taxation system, and more public initiatives".

New challenges, such as the introduction of new technologies on a massive scale, remain a constant concern to the trade union organisations, and they will continue to reduce any negative consequences of such change through collective bargaining machinery.

Turning to the attitude of employers, the report says that new technologies are often extensively introduced without previous consultation with the workers. The same applies to restructuring, rationalisation and job transfers. However, while collective bargaining has become much tougher, the report notes many important agreements. It deplores, however, an emerging trend towards bargaining at company or plant level, which it sees as accentuating the divergencies between workers as far as working conditions are concerned.

Source: ETUI: Collective bargaining in Western Europe, 1978-79, and prospects for 1980 (Brussels), 1980, 241 pp.

Fourth General Assembly of the EMF calls for more information on restructuring of industries and for technology agreements

On 30 and 31 May 1980 the Fourth General Assembly of the European Metalworkers' Federation (EMF) was held in Copenhagen; it was attended by 180 delegates from 14 European countries representing 32 member organisations and over 7 million manual and non-manual workers in the metal industry.

The resolutions approved by the assembly set guidelines for the demands to be made by its member organisations.

Noting the threats to employment posed by the economic crisis and international competitition and the permanent character which these threats are tending to assume, the EMF recommends that the struggle be intensified for the maintenance of acquired social rights, and the defence of the workers' purchasing power, by means such as new classification and remuneration systems which are not vulnerable to cyclical fluctuations in production and sales, technical innovations or changes in work organisation. The EMF also demands a 10% reduction in working hours without loss of wages, and various other advantages such as the introduction of a fifth shift, the lowering of the retirement age, the compulsory limitation of output levels by agreement and the development of apprenticeship and training.

In the light of technical innovations (in particular computerisation in offices and factories) the trade unions must stress the importance of their being associated with the preparation of projects at the design stage. Trade union delegates should enjoy the same rights of supervision and information as the management of undertakings. Technological agreements should guarantee workers retraining opportunities as well as protection against downgrading of their jobs and dismissal.

Trade unions also need to be informed at the European level of restructuring plans envisaged by European multinational companies when such plans seriously jeopardise employment (see in this connection the article on the restructuring of Philips p. .. below).

Finally, in a resolution on European policy, the EMF notes that the deterioration in economic and social policy makes it more necessary then ever to pursue a coherent industrial policy at European level. Europe must equip itself with the means to create a real, common industrial area both for the sectors in difficulty and for rapid growth sectors. The countries of Europe must take up the challenge of the third industrial revolution and avoid competition characterised by interventionism. The EMF would like to see cuts in expenditure under the Common Agricultural Policy, with the financial resources released being allocated to the European Social Fund and the European Regional Development Fund in order to increase community means of intervention for the creation of jobs.

Source: EMF: Secretariat's report on activities, 1977-1980, and resolutions presented at the 4th General Assembly, Copenhagen, 30-31 May 1980 (Brussels), Feb. 1980, 119 pp., and annexes and resolutions adopted.
Intersocial (Paris), No. 61, June 1980, pp. 21-22.

INTERNATIONAL

Bank workers and new technology: collective bargaining trends

The implications for workers in the banking industry following the introduction of new computerised technology are spelt out in a report published by the International Federation of Commercial, Clerical, Professional and Technical Employees (FIET), representing 7 million white-collar workers, including over one million members in banking. Findings are based on collective bargaining experience in 12 countries.

The report gives a number of warnings: job content and career patterns and prospects will change; some jobs will disappear entirely and over-all job opportunities will decrease in the next decade. This pessimistic view on employment prospects is shared without exception by all the unions surveyed. The service industries, of which banking is a major constituent, can no longer be viewed, as they were in the past, as a bottomless pit of employment opportunities.

The report calls for meaningful negotiations with trade unions on technological change and FIET affiliates will intensify their drive for negotiation rights on technological change, particularly in multinational banks.

Source: FIET: Bank workers and new technology - an international trade union response (Geneva), Aug. 1980, 42 pp.

AUSTRALIA

Postal workers call for benefits from technological change

The 1979 programme of the Australian Postal and Telecommunications Union (APTU), approved by its annual conference, includes a demand for the union to be supplied annually with a 10-year plan on all proposed technological changes anticipated by the Australian postal and telecommunications services. Moreover, the acceptance of technological change by the membership will be dependent upon receiving specific guarantees that benefits flowing from such changes will maintain staffing levels, improve working conditions and lead to increased leisure time. (For a recent agreement in the Australian telecommunications industry, see S.L.B. 1/79, p. 16.)

Other major conference decisions endorsed an educational campaign on the APTU Working Women's Charter, an occupational health and safety policy urging the immediate introduction of codes of practice, the setting-up of joint union/management committees to deal with training and promotions, and the extension of paternity leave.

Source: APTU: The Communication Worker (Melbourne), Mar. 1979, pp. 8-11.

Unions seek powerful action on technological change

Federal unions showed their anxiety about technological change at a conference organised by the Australian Congress of Trade Unions. Twenty unions described the effects technological changes are having in each of their areas. The conference as a whole drew attention to the increasing use of new technology systems engineering to impose: worker subservience to the machine; built-in surveillance; time and speed control over the worker; fragmentation and greater division of labour including mental labour; deskilling; dehumanisation of relations between people at the workplace; inadequate safety and health protection; low standard ergonomics. Delegates stressed the importance for unions to be consulted about the large-scale application of new technology and to ensure that its introduction brings genuine job and social benefits. It is of no consolation to the individual worker that he ends up with no job at all or one that is totally unrewarding.

Source: Australian Postal and Telecommunications Union: The Communication Worker (Melbourne), Vol. 11, No. 1, Mar. 1981, pp. 17-18.

CANADA

Trade union attitudes to technological change

It cannot be said that unions have held back the introduction of new technology, nor that workers have deliberately underutilised new technology to minimise any adverse effects on employment. So says one of the major findings of a federal government-funded study of union attitudes to new technology.

The evidence suggests that unions have not provided the degree of employment security for their members that is commonly assumed. Generally, the attitudes of unions towards technological changes is a reflection of the employment situation in the industry concerned. Attitudes are positive where there is expanding employment and negative where employment is declining. Indifference prevails amongst unions unaffected by such changes. Whether positive, negative, or indifferent, the general pattern of union reactions was reactive, not anticipatory.

Considering that over the next decade electronic technology is likely to have a substantial impact on office employment, it was surprising, says the report, that an attitude of indifference was shown by most public service organisations.

The anticipated acceleration in technological changes over the next ten years calls for more effective action on all fronts. Existing legislation and current collective bargaining provisions cannot cope with changes involving substantial manpower adjustments. All the provinces, says the report, should introduce basic regulations regarding approaches to the introduction of major technological changes, including consultation between both sides of industry before implementation. Similarly, unions and management should negotiate standing arrangements on employment guarantees, retraining, transfers and other provisions.

An examination of collective agreements reveals very limited provisions regarding technological changes. Only one-third of major collective agreements contain provisions for retraining, less than one-third provide some form of wage/employment guarantee, and only 47% provide for severance pay. The limited commitments by industry to retrain and retain mean that the public sector must develop retraining and placement policies and programmes. If the bargaining process does not evolve satisfactory standing procedures within the next few years, the legislative process will have to set general guidelines.

Most of the union members consulted felt that current

provisions for early retirement and severance pay are not adequate, even for workers who have worked as many as 30 or more years. It was suggested that economic justice dictates the introduction of policies which would guarantee a reasonable standard of living for all who have allocated 30 or more years of their lives to continuous participation in economic activity. A significant proportion of responding unions expect future technological changes will have negative effects on employment in their industries. This has important implications for labour-management relations and for public policy on the issue.

Source: Canada: Department of Industry, Trade and Commerce, Technology Branch: The attitude of trade unions towards technological changes (Ottawa), Apr. 1980, 82 pp.

DENMARK

Insurance workers adopt action programme on computerisation

At its 1978 Congress, the Danish insurance workers' union (DFL) approved an action programme on electronic data processing. Based on a report on computerisation in Danish insurance, the programme calls for sustained union vigilance and aims at securing members' full employment, meaningful jobs and a good working environment through special agreements on EDP, an EDP information service, and systematic vocational training. The Congress was addressed by the Vice-President of EURO-FIET Insurance and Social Insurance Workers' Committee, who emphasised that unions should combat the negative impact of technological change through intensified international action: control instruments should be adopted within the UN, the ILO, EEC and OECD.

Source: FIET: Newsletter (Geneva), No. 12, Dec. 1978, p. 3.

FRANCE

CFDT symposium on the tertiary sector and trade unionism

A symposium on the tertiary sector was held at Paris from 27 to 29 March 1979 by the CFDT trade union confederation, with the participation of federations in the banking, chemical, service, health, finance and postal sectors and the Paris branch of the CFDT.

The subjects discussed were (a) the definition of tertiary work (Is it - a service, a term used for administrative convenience, work in commercial or non-commercial enterprises, jobs concentrated in the tertiary sector?); (b) tertiary employment and the development of technology; (c) inter-relationship between private enterprise, government and trade union structures; (d) division of labour, labour disputes and trade union action; and (e) structures of negotiation and representation in the public and private sectors.

The trade union movement has come to have a large stake in the tertiary sector not only because of its numerical size (over 55% of the active population in France is employed in this sector - 10,748,400 persons in 1975 as compared with 6,168,000 in industry), but also because of its importance for the country's economy, and the significant economic, technological, legal and social changes that have been taking place in the last fifteen years or so. Faced with these developments the trade union movement has fallen behind in terms of trade union practices, organisation and concepts, mainly because of the polarisation of the workforce into distinct skill categories and the rigidity of collective bargaining structures. Foreseeable technological developments will tend to consolidate and even accelerate these changes. It is clear from the development of services and technological advances that the tertiary sector can no longer be limited to office work alone.

The stake of the trade union movement in the tertiary sector also has to do with the structural changes certain to occur in the sector in the near future. Three-quarters of the jobs created in France in the last ten years have been in the tertiary sector; but recent advances in computerisation radically altering job content and working methods are likely to have serious effects on the level and structure of employment.

Lastly, trade unions are keeping under close review the development of the female labour force in this sector. At present 70% of working women are in the tertiary sector, and in some branches women account for as much as 80% of the workforce; most of them are assigned to lower-grade jobs.

At the end of the symposium, it appeared that there had been "more questions than answers". However, the meeting enabled CFDT member unions to look deeper into specific features of the tertiary sector and the new problems facing it and to make headway in determining appropriate means of action and structures for representing workers in the sector more effectively.

In late 1979 the CFDT will bring out a publication on trade unionism and the development of the tertiary sector, summarising the work of the symposium.

Source: CFDT: Syndicalisme (Paris), 5 Apr. 1979, p. 2.
CFDT: Tertiare et syndicalisme (Paris, 27-29 Mar. 1979). Working papers prepared for the meeting.

CFDT: executives combat "technological laissez-faire"

This was the theme discussed by several hundred members of the CFDT affiliated Confederal Union of Engineering, Professional and Executive Staffs (UCC) on 5 and 6 December 1980 at a meeting in Paris. The aim was to draw attention to the problem of technological changes, to provoke discussion by comparing current militant experience and to show the urgent need for trade union action in this sphere. More than 50 speakers addressed the meeting, including executive-level trade unionists, sociologists, economists, engineers and ergonomists. The range of topics discussed was considerable: technological change in production workshops, offices and telecommunications, developments in the field of energy and their effects on daily life, bio-technologies and their applications to health and the environment, the international scope of technological change, the transfer of technologies and the international division of labour, and the role of the new technologies in education and training.

There was unanimous agreement concerning the effects of technological change on the number of jobs: job dislocation has been and will continue to be a major consequence, although the various categories of labour and sectors of the economy will be affected in varying degrees.

Increasing automation is greatly affecting working conditions. Although certain types of heavy work will disappear as the use of robots increases, other problems connected with mental fatigue or visual and nervous disorders may well occur.

Computer terminals which allow a person to work at home will perhaps bring about a reduction in travel time and lead to a greater flexibility in working hours; however, it is also likely to result in the isolation of the worker and a reduction in direct personal contacts.

As computers become increasingly used in production workshops and the tertiary sector, the choice of equipment will have repercussions on the choice of work organisation, which in turn will affect occupational qualifications. Whilst this may lead to a lowering of job requirements for certain highly skilled work positions (topographer, tool maker, industrial designer), other highly skilled jobs will begin to emerge in the computer field (software, programming).

As far as executive staff are concerned, although their technical role is declining as their hierarchical importance as team organisers increases, their jobs are also becoming over-qualified inasmuch as it is they who make the key decisions and devise the systems to be introduced. To the extent that their status, role and working conditions are conditioned by the effects of technological changes, the managerial staffs themselves are responsible for both pioneering and promoting these same changes.

The UCC recommends trade union action at the following levels to cope with these new technologies:

- the over-all design of the technology. This requires critical thought on the part of the trade union officials so as to anticipate management option and propose alternatives;
- conditions governing the introduction of technology. The individual and, above all, collective capacity of the workers depends upon the rate at which the technology is introduced and the range of training to be provided;
- the socio-economic consequences of the choice of technologies, which are often masked by arguments of a strictly technical nature.

Workers must be capable of devising alternative proposals at each of these levels. This in turn implies that they are provided with adequate information, that discussions are held in all the trade union bodies, that the works councils make every endeavour to monitor computer investments, that workers are given the opportunity to discuss their working conditions and in general a broadening of the scope and methods of trade union action.

This meeting was the first public event in a one-year campaign which has been launched to enable executive and engineering staffs to explore ways in which they might be able to counter the spirit of "technological laissez-faire". It will be followed in 1981 by local meetings held throughout the country, research seminars on such important subjects as robotics, electrical office equipment, etc., as well as company-level meetings to call the attention of all executives to the problem. The campaign will be closed by a meeting held in December 1981 to review the conclusions reached.

Source: CFDT: La question technologique, Cadres CFDT (Paris), No. 295, Sept.-Oct. 1980; Action syndicale et technologies, Proceedings of the UCC-CFDT Colloquium of 5 and 6 December 1980, ibid., No. 297, Feb.-Apr. 1981.
Information provided by the representative at the colloquium.

CGC report on the challenges of the new technologies

In January 1981, the Commission on Computerisation and Individual Privacy set up by the General Confederation of Executive Staffs (CGC) published the findings of eight months' work on the foreseeable consequences of the computerisation of society.

The term "novotique" was recently coined in France to embrace four new technologies: computerisation, electronic office equipment, robotics and telematics, all of which use increasingly miniaturised arithmetical and logical units.

Forty-three proposals attempt to pinpoint the developments which these new technologies will involve for industry, employment, the quality of life and individual privacy.

The chapter devoted to the impact of these technologies on industry starts with a warning. Having lagged considerably behind, France and Europe now depend on foreign markets for electronic components (90% of the EC requirements for integrated circuits is imported), and supply problems may have consequences as disastrous for the European economies as an oil embargo.

Such dependence can be broken only through a resolute Community policy backed up at the national level by co-ordinated planning (government, employers, trade unions). Such a policy will require new thinking about the sector and investments that cannot be expected to pay off for five years.

With the exception of the software industry (computer service and programming companies) where it is wellplaced internationally, France imports too much and lacks new ideas. It needs to invest on a larger scale in four swiftly developing fields: robotics (computer-assisted design and manufacture), peripheral equipment (mass memories, new magnetic disks, laser printers, etc.), teleprocessing services and data banks.

In the field of employment, the CGC recognises that the new technologies will create jobs (in 1979, staff shortages in the sector were estimated at 10,000 persons, a shortage which is increasing by 2,000 annually) but that it will also do away with many (200,000 by 1985 according to the Eighth Plan). Since automation tends both to eliminate manpower and to create new needs, a well-defined training policy is necessary if the end result is not to be a negative one.

The rescheduling of working hours and the reduction in working time brought about by the new technologies should also help to reduce the rigid structures detrimental to employment.

The development of these technologies will also have a great impact on job patterns. Generally speaking, a bipolarisation will tend to develop between high-skill and low-skill jobs. The new technologies will also affect some highly skilled jobs (millers, fitters, tool makers), which will be converted into supervisory functions, assigned at times to specialised companies.

By limiting direct operation, automation will require more maintenance staff having a higher level of skills than production crews. The proportion of supervisory staff will also change, growing in relation to the total workforce.

Restructuring will therefore be essential but cannot take place until appropriate training and higher skills are developed.

Working conditions will also undergo far-reaching

changes. Automatic operations, by removing workers from dangerous areas, will facilitate an improvement in occupational safety and health. On the other hand, the monotonous nature of some jobs (data coding, checking, acquisition), the depersonalisation of the work and the isolation of the operator faced with the machine (especially with the development of telematics) are harmful effects that must not be underestimated.

The guiding principles for the future, inevitable development of the new technologies should be consultation, training and a more realistic approach.

Source: CGC: La novotique, pour relever les défis (Paris), 1981, 155 pp.
CGC: Cadres et maîtrise (Paris), Mar. 1981.
Le Monde (Paris), 29 Jan. 1981.

FEDERAL REPUBLIC OF GERMANY

Postal union's conference on electronics and work

A two-day meeting to examine the impact of new technologies on employment was convened by the DPG - the Postal Workers Union - in Bonn on 1-2 March 1979. Special emphasis was laid on the progress of electronics in the information sector, i.e. not only in the traditional information transmission services of the Federal Post Office, but also in its telecommunications industry and office and administrative services.

Considering that the consequences of computerisation are still unclear, the DPG asked the Ruhr University of Bochum to carry out a survey on the impact on employment of electronic mail, electronic facsimile services and word processing. The survey foresees restructuring in this sector but, at the same time, new job creation: employment lost in the traditional mail services - about 1,920 jobs per year between 1980-1990 - will be partially compensated by some 1,240 new jobs per year by 1990 in the telecommunications service. Similarly, electronic facsimile services could increase employment in the Federal Post Office from 475,000 in 1985 to 489,000 in 1990. The survey indicates that the Federal Post Office employed 411,000 workers in 1961 and 462,000 in 1977, its share of total employment being respectively 1.54% and 1.84%, possibly reaching 2.14% in 1990, while the postal services' share of employment in the transport and communication sector increased at the same dates by 28% and 30.96% respectively, and may be expected to rise by 35.25% by 1990 (515,000 employees in 1985 and 540,000 in 1990).

The DPG demands that the Federal Post Office should use its statutory right to produce and sell telecommunications equipment and, as regards the electronic mail and facsimile service, it should guarantee that no private commercial interests should prevail in the provision of equipment and services. (The Post Office is one of the few in the world to be highly profitable, with over 31.8 thousand million DM turnover in 1977 and 1.98 thousand million DM profits, mainly from the telephone and telecommunications sector; 1 US dollar = 1.90 DM.)

While the DPG is not opposed to technological change, it demands that state policy should aim at restoring full employment and hence support job-creating innovations and growth-generating technologies. Technological change must be conducive to more human working conditions, especially shorter working hours for workers in hazardous jobs.

Source: DPG: Deutsche Post (Frankfurt-am-Main), No. 6, 20 Mar. 1979, pp. 3 and 16-18; and papers submitted to the meeting.
DPG: Gewerkschaftliche Praxis, Nos. 4-5, Apr.-May 1979, pp. 3, 6-10, 11-19 and a 10-page supplement.

Employee union claim concerning work environment in text processing

The Female Employee's Section of the Federal Executive of the German Union of Salaried Employees (DAG) recently published a leaflet containing guidelines for the introduction of text processing equipment in offices.

The authors note that about 2 million women work in the Federal Republic of Germany as secretaries, shorthand and typists, generally considered as a "woman's job".

According to DAG, the 30,000 word-processors introduced in offices so far have created jobs with high stress components, resembling work at the assembly line. Apart from increased health risks, this development leads to deskilling.

Moreover, DAG feels that the expansion of text processing equipment and micro-processors in the office will seriously affect the employment situation. Women will be the first to be affected: although they represent only 45% of the workforce in the office sector, they already make up 70% of all the unemployed office workers and this figure does not take into account those who have given up looking for a job.

The DAG is therefore protecting its members against rationalisation measures by taking a much firmer stand with regard to the various changes in job content and to increases in workload and stress. The issues at stake are better working conditions and environment.

The leaflet sets out detailed guidelines to reduce stress, monotony and deskilling and to improve both communication and the work environment. The guidelines range from planning the installation of word-processing equipment, work organisation and job content to health and safety measures, skill improvement as well as career and promotion possibilities. They spell out the tasks and functions of shop stewards and staff representatives concerning their participation in planning, in setting up bonus systems, fixing working hours, training and skill improvement.

A model agreement on the introduction of electronic data processing equipment and the inspection of the work environment is included, together with instructions on health and safety measures for staff working with visual display units (VDUs).

Excerpts from relevant legislation spell out the rights of workers concerning information, co-determination rights, reclassification, the work environment and rest periods. The recommendations on office workplace design and layout issued by the Federation of Mutual and Occupational Injury Insurance Institutions are also reproduced in full.

Source: DAG: Bundesvorstand Vorstandsabteilung Weibliche Angestellte: Fliessbandarbeit im Buro? DAG-Forderungen zur Arbeitsplatzgestaltung in der Textverarbeitung, 1979, 51 pp.

IRELAND

Union follows up on technology implications

As a follow-up to a major resolution passed at its 1979 annual conference, the Irish Congress of Trade Unions (ICTU) set up a Committee on New Technology. As a first step towards the development of a trade union response to new technology, a two-day seminar was organised on the "Industrial Relations Implications of New Technology" from 26 to 27 March 1980.

The seminar dealt mainly with the practical problems that will arise for trade unions in the negotiation of technology agreements.

The key issue which emerged from the discussions was that trade unions must be involved at all levels, and that new technology must be introduced on an agreed base.

The challenge for trade unions was to ensure that the impact of new technology in the workplace will lead to new employment opportunities, a better working environment and an equitable distribution of the increased wealth which will result from the use of this technology.

Source: ICTU: Report on the Seminar on the Industrial Relations Implications of New Technology (Dublin), 2 pp.

JAPAN

Union policy in the telecommunications industry

During the past 25 years the public telecommunications service has developed from a purely telephone/telegraph service to include data transmission and visual communication services. At the same time a monopoly service has become a competitive one.

Faced with the prospect of rationalisation, business intensification and unmanned operations, the Telecommunications Workers Union of Japan (ZENDENTSU) got agreement, as far back as 1965, on prior consultation before introducing new technology.

By 1972, faced with the introduction of electronic switching equipment for big exchanges (40,000 subscribers), further agreement was reached, including the following additional provisions: (1) a slow-down in the original changeover schedule; (2) written manning standards; (3) higher wages; (4) the elimination of monotonous work; (5) guaranteed employment for middle-aged workers (who might have difficulties in learning new technology); (6) 52-day company training courses following the switchover; (7) observation of the construction process by the workers so as to help them understand the new system; (8) local maintenance crews at each exchange.

By 1978 electronic switching was extended and even covered smaller exchanges (10,000 subscribers). New modifications included: (1) maintenance of existing manned exchanges; (2) when an unmanned exchange increased its subscribers to 30,000 it would be placed under the manned maintenance system.

The current collective agreement which runs to 1982, between the union and the public telecommunications service includes the opening up of new services to maintain employment. These services cover a public facsimile service; a credit card system for telephone calls, and a business information and market research service. Moreover, to improve the job satisfaction of maintenance workers more importance has been put on preventive maintenance.

Further developments planned for the 1980s include a car telephone service; experimental introduction of a video response system; a digital network for data communication; and cordless telephone services.

Source: PTTI Studies (Geneva), No. 25, winter 1979, pp. 19-25.

NEW ZEALAND

Postal workers point to social issues of new technology

The social implications of the new technologies which could be introduced into the Post Office emerge as a major preoccupation of a recent report by the New Zealand Post Office Union (NZPOU). While these new technologies present opportunities for increased efficiency and productivity, their impact on future social structures is relatively unknown.

It is "disconcerting", says the report, that so little research has been undertaken on the likely effects of the all-embracing microprocessor on the New Zealand workforce and economy.

Pointing out that it is short-sighted to gamble that new technology will somehow produce wealth and work for everyone, the NZPOU union has drawn up a series of recommendations for coping with new technology. These include the suggestion of a social impact inquiry by the Government before any major new technology is introduced, a reduced work week and shorter working life.

As far as the Post Office Union is concerned the criteria for changing over to new technology should be both socially acceptable and commercially profitable. There must be joint consultation at all stages of planning, development and implementation. Introduction should be phased over a period of time, and accompanied by retraining and redeployment arrangements for those whose jobs are affected. The Post Office must go ahead and provide new postal, banking and telecommunications services in order to generate new job opportunities for employees.

Following a review of overseas practice and individual technologies, the report is adamant that the Government has the responsibility to see that new technologies are considered in relation to their future effect on society.

Source: New Zealand Post Office Union: Technology study report 1980, 47 pp.

NORWAY

Postal and banking unions envisage joint action on automation

The Norwegian Postal Workers' Organisation and the Norwegian Bank Employees Union have decided to closely co-operate on issues related to the introduction of automation in the financial services (i.e. electronic transfer of funds from one account to another, credit card systems, etc.). The two unions organise some 32,000 employees in financial services who handle the majority of transactions in the country.

The unions have declared their opposition to the introduction of automatic systems of funds transfers and to the extension of bankautomats. They want (i) to ensure that competition between the two sectors does not determine the introduction of new technology; (ii) to be able to control technology so as to protect the respective interests of customers and employees; (iii) to ensure that employment levels are maintained and that new and meaningful jobs are created in the postal and banking sectors.

Source: PTTI News (Geneva), No. 7, July 1980, p. 6.

SWEDEN

Postal union takes up position on mechanisation

The Swedish Postal Administration is seriously considering installing automatic mail-sorting equipment in a number of centres. If it goes ahead the postal section of the union Statsanstalldas Forkund (SF) has decided to demand: (1) the right to influence the choice of equipment purchased; (2) that any staff reductions will only be achieved through retirement; (3) a maximum noise level of 70 decibels which is the limit in France and Germany; (4) medical investigations of the effects of working with the new equipment and discussions on work breaks, work rotation, etc.

The SF view is that mechanisation must be seen as a technical aid, not something that will control the pace of work.

Source: PTTI News (Geneva), No. 3, Mar. 1980, p. 5.

Unions keep watchful eye on new technology

The Swedish LO Trade Union Confederation's broad thinking on the implications of new technology will be presented to the LO Congress in September in its report on data processing.

Computerisation technology, says the LO, can bring benefits to the economy, but it is important for such technology to be introduced within the constraints of an overall economic and social plan. This means co-operation between the four levels of the economy: the Government, sectoral interests, the corporations and the workforce. At the same time, the special threats represented by data processing mean that employees must be even more adamnant in their demands for a say in controlling technical progress in the future.

The report contains two main strategies. One is addressed to trade union organisations and offers advice on union demands faced with the introduction of data processing techniques - advance notice, consultation, full information on what the effects will be on employment, work environment, job organisation and training. The other deals with the need to make the community at large more aware of the implications of computerisation. Swedish industry, says the report, is undergoing radical structural changes which should not be allowed to take place outside the influence of democratic processes. Today, a small number of companies and systems analysts have excessive power in this field. More people must be equipped to contribute intelligently to the decisions that are being taken.

At the workplace itself, LO sees the most serious threat in the impoverishment of job content. People are becoming more isolated and conversations are being replaced by anonymous messages on visual display units. Employees are subjected to more supervision and, in certain cases, there is computerised measurement of individual performance.

Possibly the most worrying aspect is the massive deskilling of the workforce that is taking place. Previously skilled work is carried out by computers which are then operated by a small number of specialists.

With knowledge being standardised, there is, says the report, a danger of deskilling not only individual employees but also society as a whole.

Right to co-determination via legislation and collective agreements, better trade union control over the work environment regulations, right to work on computerisation issues during working hours without deductions from pay, right of unions to consult experts - these are just a few of the means the unions want to have at their disposal when it comes to deciding how to negotiate issues related to new technologies.

Source: LO: Facklig data politik (Jönköping), 1981, 124 pp.
LO: News (Stockholm), No. 2, May 1981.

UNITED KINGDOM

Banking union foresees separate section for computer staff and demands 28-hour week

The National Union of Bank Employees (NUBE) is considering setting up a special computer section with its own council and annual conference. Negotiations on behalf of computer staff are already largely carried out separately from negotiations for clerical and other staff. Recent developments in microelectronics have introduced special problems and there is an over-all shortage of computer staff, in particular programme and systems staff.

Moreover at its annual conference held in Glasgow on 8-11 April 1979 the NUBE decided that, in response to the situation arising from the introduction of variations in opening hours and advanced technology, a progressive negotiating strategy should be formulated to obtain a 4-day, 28-hour week (the normal working week at present in British banks is 35 hours).

The Conference also decided to change the name of the Union to "Banking, Insurance and Finance Union" (BIFU) to take account of the fact that its membership covers the entire financial sector.

Source: NUBE: statement submitted to the EURO-FIET Trade Section Committee of Bank Workers (London, 6-7 March 1979), 2 pp.
FIET: Newsletter (Geneva), No. 5, May 1979, p. 7.

Preoccupied by new technology, TUC sets up central data bank on shorter working hours

While recognising benefits such as the elimination of much arduous and heavy work, the Trades Union Congress (TUC) continues to view with concern the employment effects of new technology (see also S.L.B. 3/79, p. 230). The issue, according to the TUC General Secretary, is not whether to accept or fight new technology, but rather how to share the benefits equitably.

As part of an over-all employment strategy the opportunity presented by microelectronics must be used, not only to change working patterns in industry generally, but also to improve the quality of life by the introduction of a shorter working week without loss of pay, longer annual holidays, less arduous conditions of work, and provision for earlier retirement on an adequate pension. To this end the TUC, in a resolution adopted at its annual Congress in September 1979, called for more, not less, government involvement in industrial planning and to provide training and retraining on a massive scale. It also called for shorter working hours and more publicly financed research to assess technological and social problems.

To provide assistance to unions attempting to secure a shorter working week, the TUC is setting up a data bank on agreements which include provisions containing any reduction on working time. The TUC is strongly recommending unions to give priority to tackling the problem of overtime - particularly regular overtime - when attempting to negotiate reductions in working time.

It is paradoxical, says the TUC, that while there are 1.3 million unemployed, many full-time manual workers still work up to 10, or even more, overtime hours per week.

Source: TUC: Resolution on new technology adopted by the 1979 Congress (Blackpool), Sept. 1979, 2 pp. (mimeo). Employment and Technology - Report by the TUC General Council to the 1979 Congress (London), Sept. 1979, 71 pp.
Financial Times (London), 11 Oct. and 27 Dec. 1979.

Optical fibres offer big advantages: POEU monitors developments

In the future, telephones will be connected to local exchanges not by electricity running down expensive copper wires as it is today, but by pulses of light carried by hair-thin strands of glass called optical fibres.

Bundles of such fibres are used to make up a cable and because of the fragility of the glass strands they are wound round a steel cord and sheathed with polyethylene. Even with the steel cord, the optical fibre cable is infinitely thinner than a conventional copper cable, but it can carry an enormous number of messages. Another major advantage is that the use of light (laser beams) makes it immune from electrical interference.

Eventually, optical fibres will be cheaper than their copper counterparts and ideal for use with the totally digital, computerised electronic telecommunications which are currently being developed by all the world's major telecommunications companies. The future of optical fibres is enormous.

In the UK the Post Office intends to introduce 15 commercial lines into the national telecommunications network over the next 3 years.

With 40,000 of its members directly engaged in working on transmission and cable systems, the Post Office Engineering Union (POEU) has a direct interest in the use of optical fibre technology in the British telecommunications network. The POEU sees four main issues: (1) responsibility for work (the Post Office has already given an undertaking that the majority of the work will be done by POEU direct labour and not by private industrial suppliers); (2) skill training in new techniques for POEU staff (training courses have already been agreed); (3) the development of satisfactory jointing techniques (a variety of techniques will be evaluated); and (4) safety factors (precautions must include measures against eye damage by laser beams). Sound safety principles are the subject of discussions between the union and the Post Office.

On the whole range of issues concerning optical fibre cables, the situation will be monitored closely by a joint POEU/Post Office Committee.

Source: PTTI Studies (Geneva), No. 25, Winter 1979, pp. 26-45.

UNITED STATES

Union demonstrates against dehumanisation of jobs

A nation-wide demonstration against dehumanising job pressures was organised on 15 June 1979 by a major AFL-CIO union, the 550,000-member Communications Workers of America (CWA). The pressures include compulsory overtime, arbitrary absenteeism controls and computerised scheduling so exacting that it turns people into machines.

The protest reflects mounting concern over the impact of automation and computerisation on the lives of white-collar office workers as well as blue-collar workers on an assembly line. A recent national survey done for the US Labor Department showed increasing job dissatisfaction, especially among white-collar workers, over the past few years (see S.L.B. 2/79, p. 164). At the same time, the National Institute for Occupational Safety and Health (NIOSH) is paying increasing attention to mass psychogenic illness among blue-collar workers. Commonly known as "assembly-line hysteria", researchers are finding that stress may cause certain physical problems among a group of workers almost simultaneously. Outbreaks are linked to jobs where the worker performs the same repetitious task over and over again.

Dissatisfaction is linked not only to technological change but also to the changing structure of the workforce which now has increasing numbers of better-educated workers, young people, women and minority groups. The CWA protest, for instance, was sparked off by a union women's conference.

Source: International Herald Tribune (Zurich), 31 May and 1 June 1979.

UAW and Ford sign up to discuss technology

Up to now, American unions have concerned themselves almost exclusively with wages and benefits. How production was organised was management's job. This seems to be changing for the United Auto Workers Union (UAW). The Chrysler crisis has resulted in the UAW's president being nominated to the management board, while at Ford they have pushed for more say when new machines are introduced. With 153,000 workers throughout the industry facing indefinite layoffs due to recession, they are becoming increasingly worried about new technology and job erosion.

As a result, in addition to similar provisions as obtained at General Motors (improved wages, pensions and sickness benefits, employee stock ownership scheme, a reduction in work time, etc.), the latest Ford three-year agreement sets up a National Committee on Technological

Progress, with five representatives each from Ford and the UAW, which will meet monthly to discuss technology issues.

Moreover, the company also agreed to provide specialised technical training programmes for workers, and has undertaken not to use new technology to erode the bargaining unit by assigning work traditionally done by UAW members to outsiders. This last point is an important one for the unions who fear that their bargaining power could be weakened as computers change the nature of jobs.

At plant level a full-time union representative will also be appointed to follow any changes in technology.

The union, however, failed to get one of its demands - the right to strike over new technology - written into the agreement.

Source: Computing Europe (London), 15 and 25 Nov. 1979.
UAW: Solidarity (Detroit), 29 Oct. 1979.

Unions debate technological change

Concern about the impact of technological change in the USA coupled with confidence in the ability of the American labour movement to cope with that change were the main items of a conference organised in Washington, D.C. in June 1979 by the AFL-CIO Department for Professional Employees. The proceedings of the conference, "Silicon, Satellites and Robots - The Impact of Technological Change on the Workplace" have recently been published and they set out clearly the range of attitudes relating to the "micro-electronic revolution" currently existing in the US trade unions and Government. (This conference was briefly mentioned in S.L.B. 3/79, p. 235.)

Probably the most sober view of current technological developments was taken by Eli Ginzberg, Chairman of the National Commission for Employment Policy. Dr. Ginzberg argued that the American experience of rising living standards and employment since the war built on the foundation of rapidly changing technology might not continue. He cited the massive increased experienced in manufacturing productivity and compared it with the sluggish record of services. The biggest impact of the computer revolution, he suggested, was going to be job losses in the white collar world. Employers would have no opportunity to keep their costs under control except by trying to get rid of people in that area.

Other government speakers took a more optimistic view. From the Bureau of Labor Statistics indeed it was suggested that the dominant trend in the US economy in the 1980s would be the continued growth of white collar employment. Some industries would lose employment as a result of the continued growth of computing while others would increase substantially. However, the point was made that job losses would not always be encountered in the firms or industries where the technology was used but could occur long after a change has taken place and in an entirely different location. This obviously made planning for technological change more difficult.

Planning for change was continually stressed by the trade unionists addressing the conference, from the AFL-CIO itself and from the telecommunications, railway and print unions. Their most frequent demand was for adequate "early warning systems" so that unions were made aware of plans to introduce changes in technology right from the beginning. Other policies stressed were the need to secure reductions in the workforce by natural labour turnover (attrition) rather than by redundancy; the need to guarantee current workers' incomes; to reduce working hours and help older workers to retire early; and to provide adequate labour mobility, training and employment placement services for workers who are displaced.

All the trade unionists at the conference were agreed on the need to negotiate collective agreements giving workers the right to be consulted on technological change and to include the policy measures mentioned. Examples were quoted of collective agreements which have already included some or all of these objectives; for example, those negotiated by the International Association of Machinists and the International Federation of Professional and Technical Engineers, one a "blue collar" union and the other "white collar".

Representatives of the US Department of Labor gave a positive response for the arguments of the trade unionists and agreed, in particular, that an "early warning system" for job losses along the lines of those in some European countries was needed. Also under consideration by the US Government, it was said, are various proposals for work-sharing, although these were seen as a short term and reversible policy.

The greater participation of American labour in policies of industrial innovation and in questions relating to the transfer overseas of technology were also raised at the conference together with the quality of working life and training policies. Both the fact that the conference was organised by the AFL-CIO Professional Employees' Department and the subjects discussed at it demonstrate the keen interest of the American unions in technology and their growing involvement in the area of white collar organisation.

Source: AFL-CIO: Department for Professional Employees: "Silicon, Satellites and Robots - The Impact of Technological Change on the Workplace", Conference proceedings (Washington, D.C., June 1979), Sept. 1979, 52 pp.

Electronic mail service gets union backing

The National Association of Letter Carriers (NALC) is supporting the introduction of an electronic mail service by the United States Postal Service.

NALC considers that the introduction of electronic mail is essential to the survival of the postal service. The service, to be introduced in about a year will involve the electronic transmission of messages but with the postal services assuring printing, enveloping and delivery of the message to its final destination. An efficient service, says NALC, will attract more customers and result in a greater volume of work for letter carriers.

Source: Postal, Telegraph and Telephone International: PTTI News (Geneva), No. 1, Jan. 1981.

NORDIC COUNTRIES

Trade unions examine implications of computer technology

According to a recent report of the Nordic Council of Trade Unions, few can now doubt that microelectronics will have a great influence on industrial technology. The rate of change is arguable, but it is clear that integrated circuit technology will bring an enormous increase in automation, industrial robots and computer control over whole production systems.

But will the application of computers and microelectronics to industrial automation necessarily lead to a loss of jobs? The report foresees that in both manufacturing industry and services there will be job losses due to higher productivity; in some cases these losses will be substantial and rapid. But against this, new jobs will be created in the production, application and operation of the software systems. There will be increased demand for system operators, system programmers, terminal and data operators. Because of the high wage costs in Nordic countries, the use of such technology becomes highly profitable. However, the total effects on employment from computerisation are practically impossible to measure but most forecasts anticipate a reduction in the number employed.

The trade unions are particularly concerned by the lack of influence that working people have over developments which directly affect them. If the difficult transition period of adaptation to the new technology is to be negotiated without widespread unemployment, the trade unions must have a say in the process so as to safeguard wage earners' interests.

Finally the report touches upon the quality-of-life options and the type of society in which people want to live in. Will the stability of society be increased or reduced through the introduction of computer technology? Will the risk of electrical breakdowns, fire hazards or sabotage connected with large computer centres create problems for society? Will industry become over-dependent on spare parts and maintenance? What could this mean in case of war or occupation?

The trade union movement will have to pay increasing attention to the consequences of computerisation and to safeguarding employment in the face of automation.

Source: Nordens Fackliga Samorganisation: Rapport fran NFS' arbetsgrupp om Datafragor, 5 Oct. 1978, 65 pp.

EUROPE

Journalists and print workers discuss press mergers and the use of new techniques

Organised by the International Graphical Federation and the International Federation of Journalists a meeting was held in Berlin from 13-15 November 1978 at which 115 union representatives from the printing trade and journalists from 16 European countries discussed mergers in the press sector and the effects of introducing new technologies in the preparation of newspapers.

In a final statement the participants note that press mergers are taking place in most countries of Western Europe, crossing national boundaries and threatening the freedom of the press. To combat this tendency they demand the restoration and protection of pluralism in the field of information and opinions, particularly by imposing a statutory obligation on companies to give employees regular information about ownership, public supervision of newspaper distribution, assistance to newspapers which are less competitive financially, the promotion of alternative solutions as regards private ownership of the press, the prohibition of mergers in new electronic media (for example cable television), and internal democratisation through greater freedom of the editorial staff (and possibly incorporated in writing in the personnel regulations).

The participants also considered that the recent introduction of new technology such as electronic typesetting, into the printing trade, constitutes a further threat to freedom of information as well as a danger to employment, small and medium-sized concerns being the most vulnerable. They therefore drew up the following demands: the right of co-determination in the introduction of such technologies: prohibition of the publication of texts originating from outside the newspaper without the knowledge of the editorial staff; the use of new photocomposition equipment to be confined to skilled compositors; shiftwork on video screens to be limited to four hours with appropriate breaks; protection against new technology resulting in wage reductions and downgrading of jobs; responsibility for news content to rest with journalists and that of production processes to rest with compositors; bar journalists from keying in and setting copy directly into a terminal; reduction of the working week to 35 hours; provision of training leave and the possibility of early retirement; maintenance of existing jobs and the creation of new ones.

The trade unions taking part in the meeting will press these claims when negotiating collective agreements (for agreements at the national level in this sector see S.L.B. 2/78, pp. 142 and 145 for FRG and the Netherlands, and 2/79, p. 151 for the Netherlands).

Source: International Federation of Journalists (IFJ): En ligne directe (Brussels), No. 112/78, Dec. 1978, pp. 1-2.

Micro-electronics and employment in the 1980s

The "micro-electronic revolution" has already led to a loss of jobs in certain key industries and services throughout Western Europe, and the situation will deteriorate unless alternative policies are pursued.

This is one of the conclusions of a report published in November 1979 by the European Trade Union Institute in Brussels which will be discussed at a conference of national union representatives from different Western European countries to be held in Oslo (Norway) from 11 to 13 December 1979.

The report's central conclusions on the negative employment impact so far are that:

- the job loss effect of the new technology has been felt first of all in those industries manufacturing products in which mechanical or electromechanical elements have been replaced by micro-electronic elements, e.g. watches, cash registers, office equipment and telecommunications equipment. Increases in the output of some goods in the electronics field have been offset in labour creation terms by the reduction in the number of components in products and the labour input required for assembly and manufacture;
- the use of micro-electronics has affected both the location and the workforce of manufacturing industry; greater emphasis on components manufacturing has resulted in a switch to producers outside Europe in the United States and Japan, who already dominate the world market;
- employment has been threatened not only by the intro-

duction of new industrial processes (for instance, in printing), but also in services (employment in banking, for example, which increased in the early 1970s, is now stagnant despite a continuing increase in demand for bank services).

Looking at the 1980s, the report points out that the significant growth in clerical employment in the last 30 years in Western European countries which has continued even during the present recession, may well be reversed as a result of office automation. Moreover, because of the high level of women's employment in this sector, women's jobs may well be the hardest hit. Over-all, it is only employment in the sectors of public utilities, construction, hotels and catering and personal and social services that will not fall as a result of technological change.

As far as the impact on working conditions is concerned, the report states that one tendency is for the introduction of micro-electronics to lead to a polarisation of employment between semi-skilled operatives on the one hand and highly skilled technical staff on the other.

The reduction of arduous and unpleasant work as a result of automation has to be offset against the increased stress and pressure of work observed in activities in which man/machine interaction did not previously take place. The new technology also increases the danger of surveillance and monitoring of workers whilst at work and also the central recording of personal information.

The report reviews current trade union efforts to negotiate all matters relating to the introduction of new technology at enterprise and industry level, concluding that:

- to avoid the negative aspects of technological change trade unions must be involved in its planning from the start and at all levels of policy making; this requires a positive response from management and full access to relevant information by the trade unions;
- the appropriate bargaining level and means of action will depend upon the national or sectoral industrial relations system;
- the key issue is the way in which technological change will affect existing patterns of work organisation, levels of employment and conditions of work;
- no-redundancy agreements have to be supplemented by additional strategies covering retraining to avoid downgrading of particular jobs, reductions in working time, wage levels and the quality of working life;
- trade union control is needed to ensure that the new technology will not be used to increase employee surveillance;
- if effectively harnessed, the extra resources released by technological change can be used to improve the

quality of work and life - this depends primarily upon the political will of governments;
- both from an employment and social point of view, the non-market services will need to be expanded; financing will have to come from the profits of the technology producing and using sectors;
- even if coupled with policies of qualitative growth the technological change of the 1980s must also allow a reduction in working time and an increase in leisure.

Finally, the report argues that a free market approach towards the growth of the European electronics industry will not reduce dependence on American or Japanese capital goods. Western European governments must assist the development of the European electronics industry, co-ordinating their actions at European level.

Source: European Trade Union Institute: The impact of micro-electronics on employment in Western Europe in the 1980s (Brussels), 1979, 164 pp.

PTTI statement on new technology

Trade union concern at the social implications of new technology was expressed yet again in a statement adopted by the Postal Telegraph and Telephone International (PTTI) at its 15th European Regional Conference held in Stockholm from 18 to 21 September 1979.

Emphasis was on the acceptance of new methods, provided there are adequate safeguards for working conditions and employment. These safeguards start from the general principle that the introduction of new technology should be introduced not only for commercial profit but also to improve services and working conditions. For this reason postal and telecommunications services must be under public ownership. The main provisions of the statement are outlined below:

- all aspects of technological change must be subject to negotiations between employers and the trade union representatives of the workers concerned;
- in order to make the choice most suitable to the circumstances in any country, planning and negotiation must proceed on the basis of the widest possible information;
- the newest technologies being contemplated have greater manpower saving potential than anything that has gone before. In addition, private companies are trying to take over profitable parts of existing services. Measures to offset the danger of widespread redundancies include reduced overtime,

shorter working hours, longer holidays;
- surplus manpower should be retained for other jobs and mobility allowances be given in cases of transfer;
- job monotony so often associated with new technology should be counteracted (choice of equipment which allows operatives to control work pace, work rotation, etc.);
- health, safety and workplace environment are subjects which require further research and continuous action;
- as computer prices drop dramatically the number of people affected by technological change will greatly increase. Training must be provided to allow workers to change from one specialised field to another and between non-technical and technical positions;
- automatic dialling, electronic switching, new means and methods of transmission will reduce the existing levels of manpower required for operating, installation and maintenance, compared to traffic handled;
- employment levels in telecommunications will in the future depend upon expanding existing services;- many of the new services made possible by new technology, such as cable television, data transmission, facsimile transmission, etc. are considered to be highly profitable by private companies who would like to enter the market and reduce existing postal and telegraph services to the role of providing only the carrier networks. In the view of the PTTI, such splintering of the industry is unnecessary. Telecommunications monopolies must be preserved and extended to cover the new services created by new technology.

Source: PTTI Studies (Geneva), winter 1979, pp. 5-12.

INTERNATIONAL

Trade Union Seminar on the Social Consequences of Automation and Technological Advance in Rail Transport

From 21 to 23 October 1974, trade unions representing 8.5 million railwaymen met in Frankfurt/Main (Federal Republic of Germany) to discuss the social consequences of modernisation and technological advance in rail transport. The trade unions represented at the Seminar were from Australia, Belgium, Bulgaria, Cuba, Czechoslovakia, Denmark, Finland, France, German Democratic Republic, Federal Republic of Germany, Hungary, India, Ireland, Italy, Japan, Democratic People's Republic of Korea, Luxembourg, Malaysia, Netherlands, Norway, Poland,

Portugal, Romania, Syria, Turkey, USSR, United Kingdom and Yugoslavia.

The secretariats of the International Federation of Trade Unions of Transport Workers (FIOST, affiliated to the World Confederation of Labour), the International Transport Workers' Federation (ITF, affiliated to the International Confederation of Free Trade Unions), and the Trade Unions International of Transport Workers (Transport TUI, affiliated to the World Federation of Trade Unions) were represented.

The Seminar called for urgent action by the International Labour Office to promote renewed discussion of the social problems arising out of the modernisation and automation of rail transport; such discussion should enable progress to be made towards the adoption of an international instrument to safeguard the basic rights of the workers concerned. The participants also expressed the view that the social consequences of the scientific and technical revolution varied according to the different social systems, and that the introduction of new techniques should be accomplished in consultation with the trade unions so as to prevent the new techniques from leading to a worsening of working conditions.

The participants noted that in recent years there has been an unprecedented decline in the level of employment in practically all national rail networks in Western Europe, and that this decline has been due not so much to technological advance but rather to government rationalisation measures which severely reduced the railways' share of the transport market at a time of rapid expansion of that market. For instance, passenger and goods transport facilities were reduced or eliminated in certain areas outside large urban centres while congestion on the highways was increasing, accompanied by a deterioration in the environment. The unions considered that a more clear-sighted exploitation of technological innovations should be possible as a means of expanding rail traffic. Comprehensive planning of entire railway operations is necessary so that workers can be transferred to expanding sectors, recruitment and training can be improved, and better conditions of work will result from enhanced productivity. This integrated planning, in which unions should participate, must take into account the following factors:

- the need to avoid unnecessary material or psychological hardship to workers directly affected, through long-term planning to adapt operations to trends in the available workforce, with transfers and retraining where necessary; the workers should be informed in advance of the measures proposed;
- suitable training programmes for new entrants, and continuous training throughout working life so as to keep workers abreast of all relevant developments;

- a smooth flow of comprehensive information and a continuous exchange of views between management and trade unions on all aspects of railway operations; greater industrial democracy would do much to relieve or obviate hardship and remove friction;
- an over-all improvement in conditions of employment on the railways, including the establishment of a pay structure which takes account of the higher qualifications required by modern techniques and the importance of an essential public service;
- a progressive reduction in working hours, a lowering of the age of retirement and greater flexibility as concerns early retirement;
- significant improvement in the working environment, through the progressive removal of all hazards to physical and mental health such as excessive noise, vibrations, fumes, etc.;
- security of employment without loss of earnings or career prospects.

The railwaymen's unions decided to continue their co-operation by holding further meetings. They requested the three international trade union secretariats to support them in their approaches to the ILO, to take urgent steps to promote regular discussion at the international level of current problems in this field, and to bring the policies of their respective sections into line with the decisions adopted and the viewpoints expressed at the Seminar.

Source: Memorandum on the International Trade Union Seminar on the Social Consequences of Automation and Technological Advance in Rail Transport, Frankfurt/Main, 23 Oct. 1974.

FIET sets guidelines on negative aspects of computerisation

Expressing concern on the current and likely impact of computer technology on the sector in which its members work, the International Federation of Commercial, Clerical and Technical Employees (FIET) held a Conference on Computers and Work in Velm (Austria), on 17-18 November 1978. The Conference was attended by 30 representatives from 21 trade unions in 12 countries together with experts and leading academics in the field of computer technology. The purpose of the Conference was to devise an international trade union strategy to minimise the negative aspects of computerisation on the quality of working life, and to maximise the use of the new technology from a human and social point of view.

A background paper assessed the likely impact of computerisation on employees in sectors covered by FIET.

Beginning with a discussion of the growth in the use of computer technology, the paper concludes that it is increasingly economically feasible to expand the use of electronic technology beyond the areas which have been traditionally suitable for computerisation. This will have a disproportionate impact on sectors within FIET's competence such as office work, banking, insurance and commerce. A particularly important development concerns "word processors" which have immense consequences for office work, printing and publishing. In addition, the signs are that they are the start of a communications revolution which will challenge existing methods of telecommunications and conventional mail.

One estimate on the employment effects of these developments has been made in a report, commissioned by the French Government, which predicted that the banking and insurance sectors alone would lose 30% of their employees over the next 10 years (see S.L.B. 3/78, p. 207 and above p. 8). FIET itself forecasts the disappearance of 5 million secretarial jobs over the same period.

Such assessments cannot be isolated from the over-all macro-economic environment: for example the development of microprocessors may open up new markets which could increase employment in some areas. However, says the FIET paper, a judgement would suggest that this employment effect is likely to be too small to significantly offset job losses. Societies will have to face the question of the distribution of the fruits of the new technology as between increased income and increased leisure.

Turning to safety and health problems, particularly visual display units (VDUs) based on cathode ray tubes, which can have adverse effects on eyesight (see S.L.B. 2/78, p. 184, 3/78, p. 277), the more general problems included an increase in repetitive jobs, reduced room for decision making, machine-paced work, shift and night work and excessive control and monitoring of performance.

The Conference called upon the FIET World Executive to draw up an action programme covering: employment (including the reduction of weekly hours of work, longer holidays, early retirement and paid educational leave to offset job losses), health and safety, job design and content, and trade union strategies. It was emphasised that control over the negative aspects of computerisation can only be achieved with strong trade unions. Trade union strategies co-ordinated at national and international levels, must therefore aim to make decisions concerning the introduction of computerisation, the result of negotiations between management and unions.

Source: FIET: Computers and Work (Geneva), Nov. 1978, 17 pp., FIET Press Release (Geneva), 17 Nov. 1978.

"Third Industrial Revolution" - metalworkers draw up a balance sheet

Delegates to the third IMF World Conference for Electrical and Electronics Industries, held in Geneva in October 1978, expressed deep concern about the future consequences of new technology on jobs and skills. Over 38 affiliated unions from 27 countries were represented by about 100 delegates who called for an active commitment by governments to research and national planning with a view to controlling the effects of technological change.

The predicted consequences of mass unemployment and skill obsolescence following the almost unlimited use of microelectronics present a serious challenge to the national and international trade union movement. Certain social scientists predict that the next generation may find it needs no more than 10% of its labour force to provide it with all the material goods needed. According to the IMF, the chip may also lead to the spread of another job-killing machine, the "robot". In Japan, already 100 producers of industrial robots are in business, compared to 20 in the rest of the world.

A further difficulty facing metalworkers is the effect of job transfer to low-wage countries. In the United States, for example, the combined effects of recession, rationalisation and job transfer has resulted in the number of workers in the electrical components industry falling from 313,500 in 1966 to 260,400 in 1978 - or by 17%. A look at some of the major companies shows the size of the problem: General Electric reduced its total workforce between 1970-77 by 4%; domestic workers, however, were reduced by 10%, whereas overseas employment went up by 27%. Phillips, between 1971 and 1976, reduced its total workforce by 13%, but overseas employment increased by 14%. Between 1970-77, Siemens reduced at home by 11% and increased abroad by 43%. Over similar periods, AEG reduced at home by 17.5% and increased abroad by 66.7%, while the figures for GEC were respectively 8.2% and 12.5%.

Among the countermeasures proposed by the Conference were: a drive to increase union membership among both manual and non-manual workers; a special effort to unionise and integrate women workers into the unions; trade union initiatives to promote policies that safeguard employment in coping with structural change; a world-wide effort to reduce working time; and the institution of special company and public funds to take care of workers displaced by rationalisation.

The Conference also studied the implications of nuclear plant construction for metalworkers and approved a scheme for co-ordinating bargaining among IMF affiliates vis-à-vis multinational companies.

Delegates examined a series of technical papers, some of which will be reported in subsequent issues of the Social and Labour Bulletin.

Source: IMF News (Geneva), No. 10, 1978, pp. 5-7. Third IMF World Conference for the Electrical and Electronics Industry (Geneva) 1978: Remarks prepared by the General Secretary, IMF.

Actors' unions confront the technological eighties

At its 11th Congress, held in Budapest (Hungary) from 25 to 29 September 1979, the International Federation of Actors (FIA), which also represents dancers, variety artists and circus performers, focussed its work on the central theme of "the actor and his union in the technological eighties". Forty-eight affiliated trade unions from over 30 countries participated in the Congress. The ILO, UNESCO and WIPO as well as the International Federation of Musicians and the International Federation of Audio-Visual Workers' Trade Unions attended as observers.

The discussions on the major theme, which resulted in the adoption of a number of resolutions, stressed the FIA's concern about the dangers facing live performance and national film and television production brought about by the increased use of new technological devices (video-discs and cassettes, cable TV, satellites); by an increased tendency by television producers to participate in co-productions with producers of other countries; and by an attitude, prevalent for example in the EEC countries, to view films not as culture but as industrial products, coupled with an increase in the use of imported TV programmes and films for national broadcasting (see S.L.B. 4/78, p. 350). These developments, according to the Congress, have had an inevitable negative impact on employment opportunities of national actors, on the protection of their rights in their works, on national culture and on artistic quality - an impact that can only worsen as the use of new technology becomes more widespread unless measures are adopted urgently at both the national and international level.

The impact of technological developments on live performance has been felt by the use of recording equipment (TV and film) in theatres. Artistically, this creates problems for the actors since the style of acting before a live audience varies considerably from that before a camera. Ballet and lyric performances are also difficult to "capture" without the camera interfering in the relationship between artist and audience. From an economic point of view the recording (or live video transmission to another hall or theatre) of the live performance presents a host of problems to actors as individuals and to the profession as a whole. Retransmission and diffusion of the recorded work (whether by television, cable TV, videogram or eventually satellite) shorten the life of the performance, placing the actors in competititon with themselves, and may replace the production of original programmes conceived specially for the new media, which affects employment opportunities for other actors. There is the added problem of how the actors can control the repeated use of their recorded works, in the absence of rights to exercise control and in the absence of a bargaining relationship with certain diffusers, in particular cable TV networks. This problem is compounded by a further problem of how actors are to be remunerated for the multiple uses of their works.

The Congress equally underscored the impact of new technology on national production of films and television programmes, which presents similar employment and protection of rights' problems to aritsts in these fields as it does to live actors. While recognising that the new technology offered some benefits and that progress in this field should not be hindered, the Congress warned that there was a great likelihood that demand for artistic activities would soon be met by a small quantity of works, produced in a few countries, that would be reproduced for different media and repeatedly retransmitted.

In the resolutions it adopted, the Congress put forward a number of measures to deal with these problems. On the issues of cinema film, live theatre and television co-production, it urged its member unions to promote in their countries a policy of aid to national cinema film industry; it urged UNESCO to proclaim an International Year of the Theatre in order to promote national policies to preserve live theatre; it declared that live performance should be protected against requirements that it accommodate itself to the demands of recording and dissemination which are fundamentally alien to the art of live performance; and it authorised FIA to negotiate with broadcasting organisations on the subject of television co-production and the dangers it presents to the employment of national actors.

On the importation of foreign television programmes, in addition to seeking guarantees of national production, the Congress warned against the use of TV programmes as "strikebreakers" and recommended that all member unions negotiate agreements with television producers and broadcasters to ensure that national or imported TV programmes are not used in the event of an industrial dispute between a union and the television producer or broadcaster in either the country of production or the importing country.

On the subject of performers' rights and new technological developments for recording and disseminating performances, a resolution was adopted calling for international and national legislation that would provide performers with rights in their performances similar to the rights provided for authors - that is, that they have an unassignable right to authorise or prohibit the secondary use of their performances. The Congress declared that such a right could be exercised in a practical and responsible way under a system of contractual licensing. It called on the international organisations to strengthen international protection along these lines.

In respect of direct satellite television, the Congress declared, as a basic FIA principle, that no affiliated union shall consider participation in this new media until it has been given guarantees of the maintenance of a level of national television production satisfactory to the unions concerned in the area of intended reception as well as in areas affected by overspill.

The Congress also urged affiliated unions to co-ordinate their efforts with other trade unions, consumer groups and public interest groups on the over-all question of new technology and on the protection for actors and the public they serve. Finally, on this issue, the Congress set up a working group to study and propose ways and means of controlling the use of technology and of ensuring performers' rights including compensation.

On other issues the Congress underscored the problems of employment, conditions of work and retraining of dancers and agreed to hold a special conference on this question. As to variety and circus artists, a resolution was adopted requesting national legislatures and the ILO, UNESCO and WIPO to intensify efforts to secure protection for these artists in the application of the Rome Convention.

The Congress also adopted a resolution to undertake studies on the actor's work and the development of the child as well as the child as artist in connection with the International Year of the Child. A resolution recommending provisions for limiting agents' commissions in agreements when fees may be payable to more than one agent in more than one country was also adopted.

Finally, the Congress adopted an Actor's Charter, embodying a set of standards on the nature of the actors' arts, their relationship to national culture, rights and duties as well as work beyond national frontiers.

Source: FIA: The actor and his union in the technological eighties, preliminary recommendation, XI Congress (Budapest, 25-29 September 1979) and resolutions adopted by the Congress, various paging.

AUSTRALIA

Postal/telecommunications workers approve future policies

The 1979 55th Annual Conference of the Australian Postal and Telecommunications Union (APTU) dealt with a wide range of subjects affecting working conditions in the industry and endorsed the following main policy discussions for future action:

- re-establishment of paternity leave provisions cancelled by legislation of 28 November 1978 (see S.L.B. 3/79, p. 266);
- the introduction of occupational safety and health courses to make staff aware of the dangers of handling asbestos, pending complete abolition of such hazardous materials;
- campaign for 30-hour week without loss of wages;
- any productivity gains from new technology to be realised in shorter working hours;
- industrial action to block the introduction of new technology unless there is prior consultation;
- establishment of a central union committee to monitor the effects of technology on members.

Source: APTU: The Communication Workers (Melbourne), Vol. 10, No. 1, Mar. 1980, pp. 7-9.

CANADA

Pregnant operators win VDU work boycott

Four pregnant employees at Bell Telephone Company (Canada) recently won the right to refuse work on visual display units (VDUs) without loss of pay.

The women, who had taken part in a work boycott, cited legislation that allows workers who believe their health is in jeopardy to refuse to carry out a particular job. Their reason for refusing was fear of radiation with relation to birth defects and the stress of not knowing whether the equipment did in fact affect reproduction.

A source of major concern was a case in August 1980 at the "Toronto Star" newspaper where four of seven children born to women working on VDUs in the same office during the same period of time had birth defects.

Although the medical evidence rejected radiation as

a positive factor in the case, there has been no substantial information explaining how that type of cluster could have occurred. The government health unit recommended that the Toronto Star hire an independent investigator to help search for a plausible explanation.

The company agreed with the union representatives that it was better to err on the side of safety. The women involved may either transfer to other work or take early maternity leave in combination with their annual leave, without losing their jobs.

Source: Toronto Star, 29 July 1980.
Computerworld (Framingham, Mass.), 20 Apr. 1981.

Unions focus on VDUs

According to the Ontario Federation of Labour, recent reports of radiation hazards posed by visual display units (VDUs) has brought about a state of alarm among VDU operators. This alarm sounded when four female employees of the Toronto Star newspaper had babies with birth defects - all in the same year although no clear cause to relationship was established (see S.L.B. 2/81, p. 132).

The processing and storing of information through this new technology has now reached an estimated level of 250,000 units in Canada. In response to the uncertainties surrounding the effects of VDUs and the lack of sufficient research, the Canadian Labour Congress has launched a major survey among 2,500 VDU operators.

Source: Ontario Federation of Labour: At the source (Don Mills), May/June 1981.

FRANCE

Terminal keyboard operators' strike ends at INSEE

A 9-week strike at the data-handling centre of the National Institute of Statistics and Economic Studies (INSEE) in Nantes ended on 16 March 1981 with the acceptance of demands put forward by the 48 terminal keyboard operators for an extra 25-minute breaks each day, the option of working one hour a day without using display screens, the suppression of individual output inspection and payment of 21 strike days.

This strike once again raises the problem of the effects of computerisation on working conditions. Until now, data input in the Nantes centre was carried out with a multiple keyboard terminal, without display screen, linked to a computer centre. Messages were coded and checked by a central computer which signalled back the errors to be corrected. The operation could thus last several months. The process was subsequently modernised by replacing the input terminal with a mini-computer equipped with a display screen which allows errors to be corrected instantaneously on-line. Although the work involved is longer and more tiring for the operators, it increases productivity by approximately 30%. Operators have complained of the visual and nervous fatigue resulting from this kind of work in which they must register between 700 and 1,000 sheets a day, involving the constant switching of the eyes to and fro from the manuscript copy placed on the worker's left side and the display screen on the right.

The local management of INSEE had earlier acknowledged the arduous nature of this work by granting two additional 5-minute breaks a day. However, the operators believed that this was not enough and on their own initiative they took an extra break of 45 minutes a day. The management reacted by deducting one-thirtieth from their wages, which in turn led to the labour dispute.

The Nantes centre is one of the terminals used for the major listings compiled by INSEE (lists of companies and establishments, electoral register, and lists of individuals).

Source: Le Monde (Paris), 13 Feb. 1981.
Liaisons sociales (Paris), No. 8460, 17 Mar. 1981.

The future of home work

A report commissioned by the Prime Minister to assess home work as a future form of work organisation and its implications for the legal and social status of workers was submitted on 16 March 1981.

The report draws a distinction between traditional home work as practised in the textile, clothing, leather, furniture, toy, clock- and watchmaking industries, and home work as it may develop owing to advances in computer communications (telematics).

The first part of the report gives a full and detailed account of traditional home work by contrasting it with the work of handicraft or factory works, citing its advantages and disadvantages for the worker (no travel, autonomy, no supervision, as against isolation, less trade union protection and insecure income), for the employer (savings on premises, supervisory staff and wages, and flexibility in staff management offset by organisational difficulties) and for the community (savings in energy and physical planning). Taking all the various factors into consideration the report was in favour of developing traditional home work to a reasonable extent, provided that the legal status and situation of homeworkers are improved.

A number of concrete recommendations are made which in practice amount to granting homeworkers the same rights as those enjoyed by workers in undertakings, particularly those rights concerning assistance in job creation, unemployment allowances, early retirement, compensation for partial unemployment, monthly payment of wages, calculation of overtime, and payment for work performed on Sundays or public holidays. When homeworkers constitute a considerable proportion of the manpower of an undertaking, they should be specially represented to ensure that their specific problems are taken into account.

Telematics - a combination of data-processing, audiovisual and telecommunication technology - has stimulated fresh thought on the concept of work and the workplace. The second part of the report describes experiments in "telework", i.e. work using telematics carried out at home or away from the main undertaking, and discusses possible developments in employment and working conditions. The experiments show that telework is still in its teething stage and will not have any actual relevance until after 1985.

The "delocalisation" of work, in particular telework, brings both advantages and disadvantages for the individual, the employer and society. A temporary solution might be the "telecentre", where a number of computer terminals and word processors are installed. Workers could work on the machines for a few hours each day and be paid at piecework rates. The machines would be connected to the central enterprise by a telephone line. This arrangement would meet with less opposition from trade unions, as it does not leave the workers isolated, does not exclude them from the working community, and gives them certain guarantees. The worker is no longer left alone and is subject to a measure of supervision. By bringing the workplace closer to the home "distance" work brings savings in time and money but maintains an industrial relationship.

Telework will also free employers from geographical limitations on recruitment, and will modify or even eliminate completely the concept of a common workplace. For the community it will simplify the revitalisation of rural areas through the installation of "lcoal communication centres".

Telework will in fact lead to the emergence of two distinct categories of workers: homeworkers and workers based in telecentres. The legal status of the latter will be no different from that of workers employed by agencies or small undertakings, as they will continue to be employees of the conventional type. Some form of teleworkers' statute should be drawn up which strengthens the worker's legal ties with the undertaking to compensate for his remoteness from it. The status of such workers should also be brought into line as far as possible with that of in-house staff and the same regulations concerning labour protection and pay should also apply.

The report's annexes include a summary of trade union views and a general outline of current home work practices in various European countries.

Source: France: Report on home work submitted to the Prime Minister by Gérard Braun, Member of Parliament, Deputy for Vosges (Paris), 16 Mar. 1981, 182 pp.

FEDERAL REPUBLIC OF GERMANY

Why display terminals in offices cause stress

A recent article, published by the trade union confederation IG Chemie-Papier-Keramik, analyses the impact on employment and conditions of work of the rapid increase in the use of visual display units (VDUs).

The changeover from physical to mental work and the related rise in office personnel costs have led to rationalisation measures in the information and data processing fields. This tendency is reflected by the spectacular increase in unemployment among office workers - a 13% rise between 1975 and 1976, representing 43% of total unemployment, and a total of 300,000 employees in May 1979, of whom 71% were women.

The rationalisation drive comprises organisational measures to control individual performance as well as computerisation of the equipment. In this context the installation of VDUs shows a rising trend (between 1977 and 1978, an increase of 30 to 40%). At present there are 300,000 VDU units in the Federal Republic of Germany, of which 26,000 are text processing units. It is expected that by 1980 banks will be using as many as 20,000 VDUs and by 1990 industry as a whole would use 1.7 million units.

This device, according to the article, not only contributes to reducing the clerical workforce, but causes health troubles (eye trouble, back-ache, exhaustion) when used for a long time. The articles refers to a research report carried out by the Institute of Labour Science of the Technical University in West Berlin, devoted to the adjustment of VDUs to the physical and psychical needs of the human being. According to this report, the human eye is stressed to its limits and often overstressed in several display tasks. The main stress factor is the frequent switching of the eyes from the copy to the keyboard and screen and back (up to 33,000 head or eye movements per day). Each eye switch requires adapting the eye to different distances and luminance, and the consequence is overstrained eyes, monotony and fatigue (on this question see also S.L.B. 2/78, p. 184 and 3/78, p. 277).

Ergonomic design of the display work unit on the basis of scientific knowledge and official standards is only one part of the solution; the other should be to avoid lowering the quality of work to the level of an inhuman activity comparable with assembly-line work. Ergonomists therefore see a need for job enrichment through diversifying work tasks and for reduced work time.

The IG Chemie-Papier-Keramik has also established a model collective agreement on work with VDUs which, besides humanisation of work, also deals with safety and health measures.

Source: IG Chemie-Papier-Keramik: Gewerschaftliche Umschau (Hanover), No. 5, Sep.-Oct. 1979, pp. 2-4.

IG Metall: pilot projects on new technology and work humanisation

Feeling that certain imbalances exist in the 6-year old state-funded humanisation work programme, the metalworkers' union - IG Metall - has started up a series of pilot projects to redress the situation. These take the form of union-run "innovation advice bureaux" and "work humanisation councils".

According to IG Metall, unions frequently have to approve collective agreements incorporating provisions protecting workers against the negative effects of technological change without being given the opportunity to influence policy before such change is introduced. At works council level, workers' representatives do not always possess the knowledge to assess work humanisation measures which, according to IG Metall, often cloak extensive "rationalisation". Representatives therefore need to be put in a better position to cope with the introduction of innovation and to judge claims relating to its impact on jobs and working conditions.

The first project, launched on 1 August 1979, set up a 3-member innovation advice bureau in Hamburg, comprising two engineers, both having wide experience in firms using new technology, and an economist/psychologist. The 3-year project is part-funded by the Federal Government (1,300,000 marks) and IG Metall is adding another 325,000 marks (1 US dollar = 1.80 DM).

The bureau (Innovationsberatungstelle) is responsible for advising workers about the effects of new technology; developing a trade union information service on the subject; consulting with companies on planned technical changes and encouraging innovation in smaller firms.

The second IG Metall project, launched on 1 January 1980 for a 4-year period, set up a 10-member humanisation of work council to give back-up information and advice to workers and union representatives, mainly at works council and supervisory board level, to help them intervene effectively, at the outset, in any projects for humanising work; to supervise the implementation of these projects and ensure that technical and organisational changes are compatible with the targets of work humanisation; to disseminate and increase the amount of information available in this area; and to help workers and union representatives to identify problems, resolve difficulties and develop work humanisation projects.

Source: IG Metall: (1) Humanisierung der Arbeit (HdA) Beratungsprojekt: Kurzinformation zum Projekt (Frankfurt-on-Main), 16 Oct. 1979, ref. 09/261/BO-H7, 3 pp.; and (2) Der Gewerkschafter, No. 5/80, pp. 10-11.
European Industrial Relations Review (London), No. 72, Jan. 1980, pp. 10-11.

Safety regulations on the use of VDUs at the workplace

The widespread use of data processing in various sectors of the economy - there are now some 300,000 visual display units (VDU) in the FRG, and their number is expected to rise by 30% per year - has prompted the joint Industrial Occupational Injury Institution in Hamburg to issue safety regulations for office jobs with VDUs which took effect on 1 January 1981. Although these regulations have the character of a recommendation they have a quasi-legal status since they constitute a commentary to the compulsory accident prevention rules.

The regulations start by listing the general standards and design characteristics regarding components of VDU work stations as laid down by the Federal Standards Institute. They then lay down specific guidelines for each component.

VDU screens must be so designed as to avoid excessive strain on operators. The most current type of screen has light letters on a dark background while the opposite arrangement (dark letters on light background) would be more suitable for the eye, since it may reduce the reflection and mirroring of the screen; the luminosity of the screen should be approximately the same as that of the printed document used by the operator while the room luminosity should be lower than that of the screen; the readability of the letters on the screen should be improved - screens should be big enough to display a sizeable amount of information at once, so as to reduce eyestrain resulting from the frequent adjustment by the eye to the information displayed; the degree of reflection and flickering should be reduced to a minimum; the letters should not "melt" together, the geometric design of letters and symbols should not be distorted and should be readable at a distance of 60 centimetres; the width and height of capital and small letters is also defined; capital letters (upper key) should only be used to convey short information, or to emphasise parts of the text.

Keyboards should be separate from VDU screens so that they can be conveniently placed and moved by the operator; the angle of incline of the keyboard should be kept at a minimum, if possible less than 15 degrees from the horizontal, and the degree of reflection on the keys should be kept at a minimum. The height of a flat keyboard should not exceed 3 centimetres.

Printed and written _texts_ which are to be computerised should be easily readable, there should be an adequate contrast between the written letters and the paper, only original texts or excellent copies should be used; no glossy paper or plastic folder should be used.

Text-holders should be ergonomically designed, the angle of the text should be between 15 and 17 degrees from the horizontal to avoid discomfort for the operator.

Tables should be large enough to allow for the proper positioning of the screen and keyboard leaving enough space for handwritten work (a minimum width of 1.2 metres, a length of 1.6 metres and a height of 72 centimetres). The table should be located at a right angle from the window and be protected from direct electric light. The operator should sit on a high adjustable and rotating _stool_.

An _eye test_ should be given to VDU operators by an authorised doctor upon recruitment and subsequently every five years; from age 45 tests should be every 3 years. When operators complain about heavy eye strain, they should see an ophthalmologist for appropriate treatment (prescription of monophocal glasses, etc.); bifocal glasses should not be used because they are likely to cause more strain as the lenses are designed for a normal reading position and are inappropriate for VDU work.

The _operators should be fully informed_ upon recruitment about the characteristics of VDU work, and its ergonomic aspects.

Source: Berufsgenossenschaft der Banken, Versicherungen, Verwaltungen freien Berufe und besonderer Unternehmen: Verwaltungs-Berufsgenossenshaft: (1) _Sicherheitsregeln für Bildshirm-Arbeitsplätze im Bürobereich_ (Hamburg), No. 10, 1980 (Ref.ZH 1/618), 29 pp.; (2) a press release; (3) Richtig sitzen und bewschwerderfrei sehen können. _Flaschen Post_, 2/80, 1 p.
European Industrial Relations Review (London), No. 86, Mar. 1981, pp. 17-18.

NETHERLANDS

FNV's call for regulations on work with visual display terminal

The Federation of Netherlands Trade Unions (FNV) urged the Minister of Social Affairs to urgently introduce legal norms concerning the use of visual display terminals (CRT), especially with a view to enabling the factory inspection to exercise control. Special attention should be paid to the global effect of radiation to which workers attending CRT's are continuously exposed. (There are currently many complaints by such workers as regards headaches, pains in the eyes, shoulders, back and neck and insomnia.)

Source: FNV News (Amsterdam), No. 8, Oct. 1978, p. 4.

SWITZERLAND

Word processing at Nestle: limited possibilities of application

During the "Computer 80" Exhibition on data-processing equipment held at Lausanne in April 1980, the head of Nestle's administrative department described several word processing experiments made by the company.

One such experiment was carried out with some 160 secretaries (for roughly 1,000 executives) divided between partially decentralised groups of three to six persons and a centralised secretariat of a dozen plus a supervisor.

All the services worked from standard texts comprising key words and basic repetitive elements. When requesting a text to be typed, they had merely to indicate the reference numbers of these various elements.

By means of this word processing system output was increased by 300%. Its possibilities of practical application, however, remain limited since, in a commercial company like Nestle, only between 5 and 10% of the total written texts can be rationalised and the gain in output is thus limited to between 2 and 6% of the total typing costs.

Another conclusion drawn from the experiment is that the advantage of a typing pool over the private secretary is, in the end, a small one. Although the output of the typing pool is quantitatively higher, the difference in the final number of pages produced is not very significant because of the greater number of corrections and retyping of pages that has to be done. A relatively high price is therefore paid for the higher output of a typing pool even though the secretaries earn less since they are no longer able to provide the personal assistance of a private secretary.

Another study, comparing the output of different automatic text-processing machines (normal golfball typewriter, automatic magnetic-card machine and full-page word processors), showed that from the strictly technical standpoint only the visual display units with large-size screens allow any appreciable increase in output (from 30 to 60%).

Source: Journal des associations patronales (Zurich), No. 22, 29 May 1980, pp. 399-400.

Ergonomic study of visual display units

As compared with conventional media (handwritten, typewritten and printed documents), the use of visual media (cathode ray tubes, microfiche readers, display units) result in specific types of eye strain. This is aggravated by postural constraints due mainly to the physical immobility of such work, resulting in static muscular activity which is much more tiring than dynamic muscular activity of comparable intensity.

Working with visual display screens (VDUs) also involves considerable mental strain and often requires new work arrangements.

In an ergonomic study of this type of work, the Labour Ecology Research Centre (ECOTRA) of the University of Geneva has devoted special attention to the eye strain involved in working with VDUs, the factors involved, and measures to reduce operator fatigue.

According to the findings, working with VDUs causes excessive eye strain when:

- the operator must keep his eyes continuously focussed on a fixed object (the screen) or constantly switch the eyes between, for example, the reference document and the screen;
- normal daylight conditions are replaced by conditions of low luminosity;
- normal reading conditions no longer exist: instead of scanning a stable surface, the eyes are forced to follow a series of successive images at the same point so that visual exploration is no longer under the operator's control but imposed;
- the flickering of the images and the colour of the screen, because of differences in luminosity with the surrounding light, results in considerable visual discomfort.

According to the study, VDU work involves three major hazards to eyesight:

- premature ageing of the organs of sight: accelerated attrition of the retina, opacity and hardening of the

lens, training the focus and eye switching mechanisms;
- acceleration of existing eye diseases whether diagnosed or not;
- aggravation of other optical defects.

Specific occupational hazards criteria, says the report, should be established for eyesight along the same lines as those already existing for hearing.

A number of recommencations are also made about the choice of screens and the job design, including in particular:

- Screen and keyboard: the display area should not be less than 17 x 23 cm and the height of the characters should be at least 3 mm. All the characters should be subject to a broad range of brightness control. The use of small and capital letters is desirable.

Text density should be average. The operator should be able to adjust character intensity and focusing. The screen should be anti-reflecting; if possible filters should be incorporated.

The keyboard should preferably be movable and self-contained. In pressing the keys the operator should not have to twist his wrists outwards.

- Lighting of the workpost: the light reflected by the areas scanned during work should be properly harmonised. Since the screen is one of the darkest areas, light from bright surfaces should be subdued by ensuring minimum illumination of the documents being used. The reflection factors and colour of the office furniture and surroundings, the layout of the workposts and the legibility of the documents should also be taken into account.

- Hours of work: for work requiring intense screen scanning or constant movement of the eys back and forth between the screen and documents, the length of work at the screen should not exceed 4 hours daily. There should also be enough breaks to allow the operator to rest his eyes thoroughly (about 15 minutes every hour and a half).

- Eye examinations should be given to test the operator's visual adaptation to VDU working conditions so that any visual deficiencies can be corrected or the working conditions themselves modified.

The examinations should be given periodically (once a year if the work involves considerable eye strain).

Source: Centre d'études des problèmes d'ecologie du travail (ECOTRA): "L'analyse ergonomique des postes de travail avec écran de visualisation", by J. Meyer, J. Crespy and P. Rey, in Cahiers ECOTRA, No. 1, IMSP publications No. 214 (Geneva), June 1980, 49 pp.

UKRAINIAN SSR

Computerised "watchdog" regulates occupational safety in agriculture

A computerised system for organising and monitoring occupational safety in agriculture was recently introduced in the Poltava region in Ukraine. The computer, using accident data received from all the farms in the region, groups them according to causes and circumstances, identifies particularly dangerous machinery and tasks and indicates the localities and jobs with the highest occupational accident rate. On the basis of pre-programmed criteria it also assesses working conditions in each farm and suggests improvements where necessary. A quarterly printout of this data is sent to local authorities and the regional trade union committee, which take the necessary preventive action.

Preliminary estimates show that since introducing this sytem, which costs only 500 roubles a year to operate (1 US dollar = 0.65 roubles), there has been a 10% decrease in the occupational accident rate (2.5 times greater than the corresponding national figure for the same period) and a reduction of some 7,500 work days lost because of accidents.

The second stage of this system, which is still being finalised, is to develop methods of forecasting and preventing occupational accidents. The system will cover agriculture throughout the Ukraine.

Source: Okhrana trouda i sotsial'noye strakhovanie (Moscow), No. 10, 1979, pp. 19-20.

UNITED KINGDOM

Office workers and electronic technology: APEX reassesses employment trends and job satisfaction

The Association of Professional Executive Clerical and Computer Staff (APEX), one of the first trade unions to analyse the effects of microelectronics (see S.L.B. 1/79, p. 7, 2/79, p. 128 and 3/79, p. 232), has brought out a further report providing a detailed analysis of developments in new office technology and the implications for employment, skill and job satisfaction among office workers.

While recognising that industry must keep abreast of new technology to remain competitive, the APEX view is that the application of microelectronics must not be allowed to go uncontrolled. APEX believes that if current practice is pursued the displacement effects of new technology will far outweigh any job creation. Moreover, young and older workers will be affected most severely. Reductions in manning levels, such as natural wastage, limit employment opportunities for those entering an already tight labour market for the first time. The Manpower Services Commission already forecasts an increase in potential youth unemployment by January 1982 (from 288,000 to 478,000). The areas where microelectronics will increase employment opportunities - computing, electronic engineering, skilled technicians - will not absorb unqualified young people.

The reduction in the number of routine manual and clerical jobs as a result of a combination of microelectronics, more sophisticated computing systems and electronic telecommunications will aggravate the situation. The main impact will be on: mail room clerks; filing clerks; clerical and administrative jobs involving large amounts of information, collation and transcription; cashiers, printing and reprographic staff; typists and secretaries; supervisors and first line managers.

APEX points to a Cambridge University study, based on a government economic model, predicting that there will be 880,000 fewer jobs in 1983 as a result of technology. The largest job loss is predicted in manufacturing industry (270,000), distribution (160,000) and the public services (150,000). Clerical and secretarial jobs will be hardest hit with 590,000 fewer jobs.

Referring to a much-publicised claim made by the Arthur D. Little consultancy that microelectronics would create 1 million extra jobs in the next decade for main producing countries (see S.L.B. 2/79, p. 128), APEX points out that, translated into figures, this only means 10,000 jobs each year in the United Kingdom.

Citing several case studies, APEX has been unable to discover a single instance where an increase in the workforce has been achieved as a result of new automatic systems. The result is always a considerable reduction in office staff.

APEX sees a partial solution in a reduction of the working week A movement towards a 30-hour, 4-day week, with a limit on systematic overtime and a simultaneous shortening of working life generally is needed if employment levels are to be protected. APEX sees this as a key issue of technology agreements which should be negotiated completely separately from annual pay negotiations.

Even so, it is estimated that only 40-75% of the net reduction in hours would be translated into increased demand for labour, and this figure may be lower if coupled with an extension of information technology.

Contrary to a recent Department of Employment report (see S.L.B. 1/80, p. 9), which sees no need for a radical restructuring of government policies to meet the onset of new technology, APEX advocates increased intervention by the Government and by the European Community to promote job creation, work sharing and better social and educational provisions for those made redundant and unemployment.

Another APEX preoccupation is that the transfer of routine tasks to a computerised system does not automatically mean more interesting work or greater job satisfaction. There is a danger that people will find themselves in more routine and less satisfying jobs. In many cases jobs are being defined in terms of the machines used rather than by the functions performed. The current use of the job title "VDU operator" fails to define the huge variety of clerical and administrative jobs which entail using a visual display unit (VDU).

The prevailing trend for computerised office systems to take over not only routine functions, but relatively complex tasks which previously provided the opportunity for some job satisfaction, will accelerate unless the pressure towards specialisation and fragmentation of work is challenged. Trade unionists must ensure that those aspects of the jew jobs which provide variety, stimulation, new knowledge, skills and responsibility must be emphasised to ensure that the majority of office workers do not become a deskilled substratum of machine minders.

Source: APEX: Automation ... and the office worker (London), Mar. 1980, 68 pp.

VDU operators code of practice

All visual display units (VDUs) and other associated systems at the International Harvester Company are now subject to a Code of practice negotiated by the Association of Professional, Executive, Clerical and Computer Staff (APEX) (see also S.L.B. 3/79, p. 233 and 4/79, p. 329).

The Code lays down standards concerning ambient lighting, reflection on the screen, frequency of flicker, character size (not less than 3mm.), control/contrast, seating, annual servicing of equipment and work breaks (10 minutes after one hour). It also ensures preliminary and regular follow-up eye examinations, with any resulting expenses paid by the company.

Source: Memorandum of Agreement between International Harvester Company of Great Britain Limited and APEX, 28 Nov. 1979, 2 pp.

VDUs - their effect on eyes?

What do visual display units (VDUs) really do to eyes? Are they really as bad as some say? Two leading opticians have made a study of the effects of working with VDUs.

Acting as consultants to both sides of industry, they join the ranks of those who agree that VDUs are not harmful to eyesight - if proper safeguards are used. These include eye tests. "Ideally", says the study, "operators should have their eyes examined prior to using VDUs, to identify any existing ocular defect". It has to be accepted that a minority of people would be unsuited for VDU work. Subsequently, it is in the interest of both employer and employee for VDU operators to have regular eye examinations. The study recommends specific visual standards for VDU operators.

By the visually exacting nature of their work, VDU operators are also more prone than average to the symptoms of eyestrain, such as tiredness, irritation, screwing up the eyes, discomfort in bright light and headaches. Here again the study suggests that movement behind the VDU should be at a minimum to avoid needless stress. Colours should be restful and temperature, lighting and humidity were important factors, especially when concentration may reduce the blink rate.

An observation noted by the authors in a survey, but not explained, is that VDU operators, even of long standing, arranged their work areas differently from staff using traditional equipment. Those on traditional equipment personalised their working area with pictures and personal possessions, while the VDU area had a clinical appearance. Discussions revealed that the VDU operators did not fully associate themselves with their work, even though it was their permanent job.

Source: Vision and VDUs by J.W. Grundy and S.G. Rosenthal, Members of the Occupational Visual Welfare Committee for the Association of Optical Practitioners (London), 1980, 13 pp.

UNITED STATES

Union coalition on VDU dangers

Six AFL-CIO North American unions and 25 local unions have formed a coalition to publicise alleged dangers of visual display units (VDUs) and to urge the government to study the effects of using such units on employees.

The unions maintain that since the equipment was introduced, employees have complained of headaches, tension, eye strain, neck and shoulder pain, and even colour blindness. One member of the coalition - the Office and Professional Employees International Union (OPEIU) - states that it is thinking much more seriously about the problem than it did previously.

The coalition has asked the National Institute on Safety and Health (NIOSH) to conduct a study of radiation emissions of VDUs. NIOSH had yet to respond to this request but it has already agreed to an 18-month exploratory study to determine if the VDUs contribute to eye and back strain.

Source: Labor Relations Reporter: News and Background Information (Washington, D.C.), Vol. 102, No. 27, 3 Dec. 1979, pp. 299-300.

How far away is the "electronic home office"?

According to Norman White of New York University's Graduate School of Business Administration, the "office of the future" is still 10-15 years away. At the first Conference on Office Automation held in Atlanta (USA) from 4 to 6 March 1980, Dr. White and fellow academicians addressing the Conference said that what is really wanted is not dramatic claims and counter-claims about technology-induced unemployment but an identification of avenues for further research into the effects of the office of the future on society in general and on office workers in particular.

While no one can predict what will happen, there is no doubt that workers will need a familiarity with and acceptance of the role of computer technology. While electronic games and home computers are leading in that direction, the real catalyst will be the educational system.

While new management functions will emerge, the manager of the future will have to improve writing techniques because, through telecommunications, he will have less face-to-face communications with employees. There will be less emphasis on appearance or the "managerial image", and certain groups, such as the handicapped and minorities, will be afforded greater career opportunities than at present.

In discussing the eventual possibility of employees working from home operating a visual display unit (XDU), the Conference agreed that the idea is not very attractive. Social interaction was seen as extremely important to the psychic well-being of the average worker. But, according to the Conference, the trend is definitely in that direction, with increased fragmentation of traditional office and, eventually, societal structures.

Source: Computer Weekly (London), 17 Mar. 1980, pp. 15-16.

Print unions act on results of official study on visual display units and operators' health

VDU operators suffer a "significantly higher percentage" of eye strain, itching of the eyes and aching shoulders than non-VDU users, says a report by the National Institute for Occupational Safety and Health (NIOSH). While operators are apparently safe from radiation dangers, which are too low to be hazardous, the report indicated that anxiety problems might be intensified.

These are the preliminary findings of an on-going survey carried out by NIOSH at the request of local and national union groups (see S.L.B. 1/80, p. 95). Tests were carried out among the employees of three leading newspapers.

Among the preliminary recommendations to counteract the adverse effects among VDU operators were: (1) regular rest breaks; (2) initial and periodical eye tests; (3) adjustable keyboards and screens (too many operators were looking down at the screen); (4) adjustment of room lighting to avoid glare; (5) testing of VDUs for radiation emissions before being used again after servicing.

Apart from corrective procedures to deal with the problem of glare, no detailed recommendations are expected before the end of the year.

Following these first findings, the Newspaper Guild has instructed its local unions to include breaks of 15 minutes after 1 hour or 30 minutes after 2 hours of VDU work in all contract negotiations. The Guild and other unions are co-operating with the NIOSH study.

Source: Computerworld (Framingham, Mass.), 2 June 1980.

Laser printers use suspected carcinogens

Five out of six laser printers examined by the Environmental Protection Agency (EPA) in the United States contain chemicals suspected of causing cancer. The main offenders are the selenium or nickel used in the photoreceptors.

The EPA, the Food and Drug Administration (FDA) and the Occupational Safety and Health Administration (OSHA) all list both these chemicals as possible carcinogens that deserve further testing.

Source: Computerworld (Framingham, Mass.), 3 Nov. 1980, p. 10.

Proposed study on visual display units

The Newspaper Guild, in co-operation with the Mount Sinai School of Medicine in New York City, plans to launch an extensive study of the health effects of visual display units (VDU).

The study should prove to be one of the most

extensive carried out so far on the effects of VDUs. As well as monitoring the effects of VDU operation on users, it will also seek information on their medical histories and previous radiation exposures.

The study will be co-ordinated with an on-going investigation by the AFL-CIO sponsored, Workers' Institute for Safety and Health into the effects of non-ionizing radiation at the workplace. Meanwhile, the International Typographical Union recommends the installation of metal shields around transformers wherever one milliwatt per square centimetre of radiation has been detected.

Source: The Bureau of National Affairs Inc.: Labor Relations Reporter, News and Background Information (Washington, D.C.), Vol. 106, No. 25, 23 Feb. 1981, pp. 156-157.

Committee says no health risks with VDUs

Operators using visual display terminals (VDUs) face no health risk from radiation emissions, according to testimony presented earlier this year to a House of Representatives science subcommittee.

This conclusion was reached after two days of controversial testimony from manufacturers, federal health agencies and newspaper publishing and trade union groups.

Several witnesses had referred to reports by the federal health agencies which supported industry claims that VDU radiation is well within safety levels.

Despite the subcommittee finding, the chairman ventured to express the opinion that the "question of possible effects, including cataracts", as well as "stress effects" continued to be on many people's minds. And, so far, no epidemiological studies or experiments with laboratory animals had been carried out to confirm the findings of federal health agencies. The subcommittee would like to see such studies carried out.

Source: Computerworld (Framingham, Mass.), 25 May 1981.

EUROPE

European banking staff discuss opening hours and rationalisation

The EURO-FIET Trade Section Committee of Bank Workers representing 550,000 in 22 countries met in London on 6-7 March 1979.

A major item on the agenda was rationalisation and its effects in reducing job opportunities and the Committee fully supported the proposed FIET Action Programme drawn up in November 1978 (see S.L.B. 1/79, p. 42) to counteract the negative effects of computerisation. The programme includes a reduction in weekly hours of work to offset job losses. Strongly supporting unions who refuse an extension of bank opening hours at a time of increased automation, the Committee stressed the need for designing national and international policies to cope with its effects.

Source: FIET: Newsletter (Geneva), No. 4, April 1979, pp. 1-2.

Rationalisation in the retail trade: some survey results

Nine trade union federations in eight European countries (Austria, Belgium, Federal Republic of Germany, Denmark, Italy, the Netherlands, Sweden and Switzerland) answered a questionnaire on rationalisation policy in the retail trade sent out in 1978 by the International Federation of Commercial, Clerical and Technical Employees (FIET).

The main findings showed an extension of self-service, increased concentration of the market in the hands of a few enterprises and centralisation of functions within companies. These structural changes in turn facilitate organisational and technical rationalisation processes.

As far as technical rationalisation is concerned, electronic data processing (EDP) is pre-eminent in all 8 countries, particularly in automatic stock control, labelling systems and electronic cash registers. Moreover, electronic cash registers are more widespread than the application of EDP to wages and salaries accounting (reported from 6 countries), invoicing (reported from 5 countries) and in the organisation of merchandise (6 countries). Jobs are directly eliminated or changed by the introduction of these technologies.

In Sales, it is anticipated that the video screen will be used increasingly, especially in the mail order trade, for information on delivery times and other sales conditions. At the same time, the introduction of text-processing systems is leading to the disappearance of a large number of clerical and secretarial jobs. In 1977, in the Federal Republic of Germany, the proportion of the unemployed of both sexes coming from the retail trade amounted to over 146,000, i.e. 16%. This figure included more than 97,000 women, representing 20% of all unemployed women in all sectors. Moreover, between 1975 and 1976, job reduction in the retail trade in Italy amounted to 12.5% (100,000 jobs), 5 to 10% in Switzerland and 2.7% in Belgium.

Repercussions. The employment effect is not the only matter for concern. The new technology raises many other issues, including questions of job satisfaction safety and health. Tasks tend to be deskilled, while, at the same time, the work tempo increases and, with it, the stress under which employees operate. Economic problems arise if the criteria originally used for assessing the rewards payable to employees (i.e. professional training, experience and responsibility) are abolished, as the deskilling of jobs automatically means lower pay.

Within the context of these findings the FIET Conference of Commercial Workers, meeting in Geneva on 29-30 November 1978, endorsed the proposed world action programme currently being drawn up by FIET (see S.L.B. 1/79, p. 43). The Conference also stressed the urgent need for control and consultative machinery to identify in advance the negative effects of technical and organisational change and to take steps to prevent their occurrence. Rationalisation should be used not only for higher profits but also to increase the humanisation of work.

Source: FIET: Effects of rationalisation on the employment situation in the European commercial sector (Geneva), 1979, 31 pp.

EMPLOYMENT, TRAINING AND QUALIFICATIONS

AUSTRALIA

Computers and job losses in the print trades

People should also get some benefits from the computer revolution says the Australian Print Union (PKIU) and the Australian Congress of Trade Unions (ACTU). The demand goes back to 1978 when PKIU, the first union in Australia to do so, got involved in a conflict over the introduction of computerisation at John Fairfax and Sons Ltd., the largest of the four newspaper publishers in the country.

The direct cause of the dispute were lay-offs caused by the terminals which enable both journalistic corrections and print-setting to be carried out at the same time. Since then the Teleprint Services (Telecom) and the telephone network have announced staff reductions of 20% and 60% respectively over the next 6 years due to the introduction of computerised systems.

The PKIU general secretary recently pointed out that while Australia for the last 15 years has normally been about 5 years behind the USA and Western Europe as far as implementation of technological developments was concerned, this has now changed, and today new printing equipment is often ordered while it is still in the development stage.

Other new technologies that have recently been introduced in the Australian print industry, including for example "ink-jet-printing" which is still in its infancy, but does away with both "hot metal" and photo matrices, have also caused concern in union circles, not least as some of them have been tried out by strike-breakers in recent years. The unions are also concerned by the fact that the number of jobs in the printing industry fell from 59,159 in 1974 to 51,675 in 1978. These are substantial figures in an Australian context and the trend seems to be continuing.

Today the PKIU is demanding shorter working hours at the video terminals, job guarantees and compensation payment in case of technological lay-offs. Their more general long-term demands include a 32-hour working week, 4 weeks vacation twice a year, and a lowering of the pensionable age to 60, protection against loss of jobs and compensation from the employers or the State in case of inevitable job loss.

Only in this way can the working people in Australia glean benefits from the computer revolution.

Source: Druck und Papier (Stuttgart), No. 9, 4 May 1981, pp. 20-21.

AUSTRIA

Siemens retrains staff in microelectronics to avoid redundancies

In a lecture given in January 1979 to members of the Austrian Institute for Education and Economics, the Director of the Austrian subsidiary of Siemens AG, the electrical appliances and electronic firm, stated that in some ways microelectronics had negative employment effects. For instance, the production of microelectronically controlled telex equipment had led to the disappearance of numerous jobs. On the other hand, increased sales, servicing and maintenance largely made up for job losses by the creation of new jobs which required different and higher qualifications.

In Austria, therefore, Siemens has developed a re-training programme in collaboration with the labour market authorities. Skilled workers made redundant by the introduction of new technologies will, within one year, be trained as microcomputer technicians. Participants return four to five times a year to their jobs in order to apply their newly acquired skills. Training techniques which have proved to be particularly suitable to adults are used, such as video-films, tutoring, practical exercises, independent study and group work.

Source: Oesterreichisches Institut für Bildung und Wirtschaft: Mitteilungen (Vienna), Jan. 1979.

Trade union meeting on microelectronics and job opportunities for girls

Microelectronics and job opportunities for girls was the theme of a meeting held in Vienna on 13 May 1981 by the Women's Section of the Austrian Metallurgy, Building and Power Supply Workers' Union (MBE) under the chairmanship of the Secretary of State. Speaking to the press, the MBE President said that microelectronics was having an increasing impact not only on the economy but also on the labour market, in particular on women. The MBE must prepare itself to deal with the situation by finding out what had been done in other countries in order to avoid mistakes already made and to improve vocational training, particularly for girls.

A representative of Siemens, the electronics firm, said that for some years his company had been successfully training girls in metalworking occupations such as precision engineering, toolmaking, communications electronics and electro-mechanics. Although girls sometimes did better than boys in theoretical subjects, they had greater difficulty in workshop practice. This was due to their education and to the role assigned to them in society and had nothing to do with their aptitudes.

An employers' representative mentioned the physical and psychological disadvantages of modern technology - sustained attention, monotony, concentration - and stressed that works councils should be informed in good time of changes and their effects in order to be able to participate in decision-making and not merely to be faced with a fait accompli.

A trade union vocational training specialist drew attention to the fact that micro-electronics would be permeating many sectors of the economy; the MBE should ensure that the new occupations were open to both girls and boys without distinction. Girls should be made to understand that whatever occupation they entered, they would probably need to have a knowledge of electronics. Without undergoing an adequate apprenticeship in a metal trade, for example, they would not get a job in this sector, nor would they receive the supplementary training required by the rapid development of technology.

Source: "Gewerkschaft MBE: Chancen der Mikroelektronik nützen", in OGB - Nachrichtendienst (Vienna), No. 2085, 21 May 1981.

CANADA

An overview of the current debate on employment effects of technology

In view of the widely conflicting claims being made about the job displacement effects of technology, the Canadian Department of Communications commissioned the Institute for Research on Public Policy to carry out a review of the most significant world literature dealing with the impact of computer/communications technology on employment. The result is some tentative observations on the nature of the debate and some conclusions about needed research.

Based on a survey of over 4 million titles - whittled down to nearly 400 relevant and 42 key documents - the report calls attention to the lack of facts supporting the current controversy which polarises people into two groups which supporting either the negative (job destruction) or positive (job creation) effects of new technology. Clearly the labour spokesmen have a tendency to stress the negative aspects, while industry stresses the positive side.

According to the report, all attempts to analyse the issue quantitatively have so far been less than satisfactory. No account is taken of the development of entirely new industries and, more important, intersectorial linkages are ignored. Moreover, the limited theoretical underpinning of the debate in general is remarkable.

The literature review reveals absolutely no consensus on the key question of the net job balance effect for the whole economy. Most of the reports however are in agreement that robotics and numerical control machines will cause displacement in manufacturing, and that word processors will invade offices and displace secretaries. There is also some agreement on job destruction in specific segments of the labour market: these are thought to be more severe for older workers and for the less skilled members of the labour force. Middle management will also suffer. But over-all, it is the women, dealing with information processing functions in the service sector, who are expected to bear the brunt of the impact.

No one really knows, says the report, what the net employment outcome of this new wave of technological change will be. While the unemployment rate in most OECD countries has doubled over the last decade, rising unemployment is not necessarily an inevitable consequence. Further monitoring of the issue on a continuous basis is essential.

Aware of the dangers of using aggregate statistics, which often obscure as much as they reveal, the Institute tried to get more detailed information on the technological and economic changes which have influenced certain sectors. Contrary to expectations, the few concrete situations examined differed radically from what the reviewed literature, based in turn on aggregate statistics, would have led readers to believe. Intensive, concrete investigations of actual technical and structural changes in the economy must therefore complement aggregated statistical approaches to analysis.

In this context, the report calls for further macro-level research along the following lines:

(1) Admitting ignorance of this extremely complex and dynamic phenomenon.
(2) Accepting that there is no inherent reason why humans cannot be replaced by technology.
(3) Constructing scenarios of the impact of microe electronics on employment (ranging from optimistic to pessimistic) each based on explicit assumptions about the diffusion rate of microelectronics, about international technology transfer, capital requirements, the social adaptation rate and other sensitive variables.
(4) The scenarios should consider the Canadian economy in a model in which all sectors of the economy are seen together. Such a model would allow the investigation of the crucial "what if" questions which are currently ignored.
(5) Finally, an interdepartmental task force should bring together technological and labour market expertise to examine the nature of the technology and its rate of diffusion.

Only this way is it possible to assess if we are really faced with the third industrial revolution or a case of traditional technological change and "business as usual".

Source: Institute for Research on Public Policy: The impacts of computer/communications on employment in Canada - an overview of current OECD debates (Montreal), Nov. 1979, 296 pp.

A potential threat for jobs in telephone manufacturing and maintenance

Allowing customers to buy (eventually from abroad) and connect their own telephone installations to the telecommunications network, instead of renting them from Canadian companies, would directly threaten jobs in Canadian telephone manufacturing and maintenance companies.

This is the main conclusion of a study undertaken by the Canadian Federation of Communication Workers (CFCW), after the Bell Canada Telephone Company referred the question to the Canadian Radio and Telecommunications Commission (CRTC).

The Federation gave particular attention to the effects of such a decision on the monopoly position of Canadian companies, on prices and on whether foreign equipment would be compatible with the existing Canadian network.

The Federation is completely against any alteration of existing legislation and points out that the threat to national companies will become extremely serious if the Government does not take adequate measures to protect them from international competition.

Source: PTTI News (Geneva), No. 7, July 1980, p. 4.

FRANCE

Employment consequences of electronics in the telephone industry

Technological changes in the postal, telephone and telecommunication (PTT) sector have already adversely affected the level of employment in both the telephone industry and in the postal services, and the trend is likely to worsen.

In the telephone and telecommunications manufacturing industry, the decision taken in 1978 by the Directorate General of Telecommunications (DGT) to change from space-switching to digital or temporal systems, which require four times less manpower, led the manufacturers (some six major companies and their subsidiaries) to consider large-scale reductions in their labour force.

The industry's trade association SITT, has said that some 15,000 jobs will disappear over the next three or four years, out of a workforce of 90,000. According to the trade unions, 2,000 workers have already been or are about to be laid off. The initial estimates indicate that the categories most affected are manual workers, whose number may be cut by two-thirds (especially women), and work site technicians. Layoffs will affect regions already hard hit by the current recession (including Brittany) where the government has made a special effort to install high technology industries. The situation has been aggravated by the relative failure of exports (a 19% turnover instead of the expected 30%) and the rapid expansion of telephone production (30% annually since 1975) which has not only made good the previous shortage but has reached a certain degree of saturation and levelled off output.

A report by the National Centre for Telecommunications Research (CNET) gives even more pessimistic figures: direct and indirect employment in the industry, which amounted to 150,000 jobs in 1977, will drop to 100,000 by 1985 and to 80,000 by 1990, even though several thousand new jobs may be created by developing new products. According to the report, these job cuts should be offset by an increase in PTT administrative staff of approximately 115,000 new jobs (252,000 employees in 1990 as compared with 137,000 in 1977) because of the development of the telephone network (30 million lines in 1990 against the current 12 million).

The Secretariat of State for the Postal, Telegraph and Telecommunications Service, which in France acts as "watchdog" for the industry and is its sole customer, considers that job-shedding fears are excessive; and that cutbacks of 10,800 will be offset by new jobs in the postal administrative services for which redundant workers in the telephone industry will receive priority: 45,000 new jobs have already been created (administration, building, civil engineering and manufacturing) and the 1978 export turnover was very encouraging. The DGT estimates that 600 manufacturing jobs will eventually be created when exports reach 30% of turnover and that an additional 8,000 jobs will come from new products (keyboard consoles, telecopying equipment, electronic telephone directories), as well as 20,000 administrative jobs by 1983 and 6,000 jobs as a result of the telecommunications satellite and its receiving stations. The Government also intends to exert pressure, especially in connection with credit, on the companies contemplating dismissals.

Source: Le Monde (Paris), 20 Mar. and 4 and 21 Apr. 1979.
CFDT: Syndicalisme (Paris), 15 Mar. 1979, pp. 4-5.
Financial Times (London), 15 Dec. 1978.

New technologies cut back jobs at Renault

According to the head of Renault's personnel department, 18 jobs out of 100 will be done away with by 1990 as a result of technological developments. At the same time, a further 6 will be created, which means that, given the same production level, job losses will amount to 12%. These changes also affect work content, and the number of assembly-line workers has dropped by 8,000 in the last 12 years, to reach 25,000 at present.

According to the personnel department, further training will help to cope with these changes, as well as a reduction in working hours (39 hours 10 minutes weekly at present) and in the total length of working life (in 1980, 2,540 persons took early retirement and it appears generally that many workers are prepared to take a cut in earnings in order to retire early). A fifth week of paid holidays, however, is not contemplated as a possible solution for the time being.

As regards the 19,000 immigrant workers employed by its various plants, Renault states that, despite technological developments, there will be no large-scale layoffs and the workforce will be trimmed by more flexible measures (for example, by three to six months' leave without pay).

Source: Le Monde (Paris), 14 Feb. 1981.

FEDERAL REPUBLIC OF GERMANY

Training needs of new technology

The impact of the widespread use of microprocessors (MP) and microcomputers (MC) in an increasing number of economic activities and on education and training policies in the Federal Republic of Germany, Japan and the USA was the subject of a Battelle Institute study sponsored by the Federal Ministry of Education and Science. Over 300 undertakings, numerous experts and affected workers provided information.

The study contains a wealth of data which underpins the conclusions regarding development tendencies in training up to the technician level. It starts out by showing that microelectronics is a key technology and gives interesting figures about the past growth of this sector, which is expected to absorb increasing numbers of workers while displacing workers in other sectors. The reasons for this growth are the drive to increase productivity, to improve the quality of products and of working conditions and environment (abolition of monotonous and dangerous work) and to enhance flexibility of production and to cope with increasing complexity of industrial processes. The study shows that the application of this new technology advances in leaps and bounds.

So far there are only rough estimates on the number of workplaces directly affected by computerisation. For the USA the figure is between 2 and 3% of all workplaces at present but it is expected to be around 40% by the year 2000. In the Federal Republic of Germany it is estimated that up to 2.5 million workers will be affected in the next five years, and in the long run about 50% of all workers will be concerned; and about 500,000 trainees are presently enrolled in apprenticeship trades whose content will be altered by the new technology. In the three countries there will be reduced employment opportunities for unskilled and semi-skilled workers. The emergence of entirely new occupations has so far been limited and changes usually take place within existing occupations. Only the MC/consultant and the MC/technician have been mentioned as new occupations.

While there will be new and changing job requirements in the microelectronics field, more general skills such as abstract thinking, planning skills and ability to work in a team are gaining even more in significance. This is why general education should put more emphasis on such general skills. The educational entry requirements for skilled workers' training in the electronics occupations are being raised as a result of the new technology. Most existing skilled workers will need to acquire supplementary qualifications in microelectronics.

In all three countries, microelectronics and computer technology are so far mainly imparted in a great variety of further training courses by institutions and undertakings. There has been little systematic effort to include it in basic education and training curricula although it is recognised as a key technology. The training systems react hesitantly. Lack of familiarity of teachers and instructors with microelectronics and industrial practice and developments has been identified as a major cause of this delay. The most consequential efforts to overcome these shortcomings have been made in Japan.

Governments in various highly industrialised countries, except the USA, have reacted to the challenge and are attempting to implement a strategy to cope with this new technology. Its elements are as follows: mondernisation and rationalisation of small- and medium-sized undertakings by promoting the application of microelectronics throughout the economy; early training and further training of specialists to overcome existing bottlenecks; inclusion of basic knowledge in data processing and microelectronics in secondary school curricula to prepare young people for future job requirements.

It is certain that this technology will change industrial societies although there are obstacles to overcome such as workers' fear of redundancy and obsolescence of qualifications, managers' ignorance of MP/MC applications, lack of risk capital and unwillingness to take risks in view of existing capital stock and investments. However, there appears to be a basic consensus that the wide application of MP/MC technology must not be hindered in the interests of economic competition and the creation and preservation of jobs with a future. The structural and individual effects would have to be overcome by adequate measures. The education and training systems are beginning to respond. The new technology reinforces the already existing trend towards continuous and lifelong learning and training. Initial education and training will only provide knowledge for starting positions in the future information society.

Source: R.V. Gizycki, U. Weiler: Mikroprozessoren und Bildungswesen (Munich, Vienna), R. Oldenbourg Verlag, 1980, 316 pp.

NETHERLANDS

Services sector unions scrutinise the impact of automation on employment

The trade, banking and insurance union of the Netherlands' biggest confederation of trade unions, FNV, published in May 1979 a brochure on the effects of automation on employment. Accepting that changes as a result of automation are a normal feature of a growing economy, the brochure points out that, in the current situation of slower economic growth, rationalisation is used as the main method of reducing costs and of improving a firm's competitiveness. As a result of automation, the volume of employment is therefore likely to decrease, making structural unemployment a more permanent feature of the economy.

Turning to the effect of automation on the quality of employment, i.e. its content and organisation, the brochure forecasts that, in general, the content of work will be less interesting and, although not inherent to the process of automation, there will be greater centralisation in the organisation of work.

The brochure contains detailed information about automation and employment in the services sector. Although the main labour-absorbing sector during the last 15 years, rationalisation in various branches means that it is now no longer capable of offering much more employment. In the retail trade, for example, there is an increasing use of cash terminals. Banks have reached the limit of efficient personalised services and are trying to make more payments automatic. Insurance companies have been somewhat later in rationalising their operations but are now centralising files for each insured person. Finally, the use of word processors is likely to reduce typing staff by 20 to 30% within 10 to 15 years.

The trade union's position as set out in the booklet recognises the importance of automation but disagrees with the way it is applied. The trade unions are particularly worried about the fact that social factors are not considered and feel this is to the detriment of both the employer and the employee. As to the volume of employment, the trade unions generally propose to increase government employment, to restructure weak enterprises and/or sectors and to reduce working time in order to distribute the volume of employment more equitably between those who want to work. A reduction in working time, including early retirement, should be negotiated at the enterprise and sectoral level, within the so-called "employment agreements".

As to the quality of employment, unions should have more say in decisions on how automation is being applied and to what extent. The discussions between management and employees should deal with the content of tasks, with experiments and with retraining. It is suggested that legislation be adopted (similar to that in some Nordic countries) which obliges management to discuss with the employees the adoption of automation and its effect on the volume and quality of employment.

Source: Dienstenbonden FNV: Automatisering en werkgelegenheid - een vakbondvisie (Automation and employment - trade union's viewpoint) by Van Gelder, W. (Woerden), May 1979, 72 pp.

Report of the Advisory Group on Micro-electronics

On 5 November 1979, Professor Rathenau, former director of Philips physics laboratory and chairman of the government-appointed Advisory Group on Micro-electronics, presented his report to the Minister of Science Policy. He was assisted by members of the university community and by representatives of the private sector in the preparation of this report.

After explaining some of the fundamentals of micro-electronics such as digital micro-electronics, technical aspects and applications, the report sketches the possible consequences of the micro-electronic revolution on the Netherlands economy. In so doing it deals with social and cultural influences, the role of education, and sets out policy recommendations.

For the ILO constituents the most relevant part of the report is the review of the estimated impact of increased application of micro-electronics on the volume and quality of employment. On the basis of a macro-economic model it is estimated that unemployment will increase by about 150,000 persons (actual level of total unemployment is about 200,000), if no corrective policies are pursued. Even if the introduction of micro-electronics were slowed down, about the same level of unemployment would result. This is due to the fact that slow introduction of microelectronics would weaken the competitive position of the private sector and would reduce exports. It is implied by the report that the unemployment problem can only be handled by a policy which aims at reducing labour costs and working time. In order to improve the competitive position of Dutch enterprises the report recommends the establishment of a micro-electronic centre which would provide advice to mainly medium- and small-scale enterprises.

The report finds it difficult to assess the impact of the micro-electronic revolution on the quality of employment (conditions of work). On the one hand, it can increase solitude, the power of a few (super-) technicians and the speed of work. The report concludes that it depends mainly on management whether positive or negative aspects will prevail.

Source: Netherlands: Maatschappelijke gevolgen van de Micro-electronica - Rapport van de adviesgroep Rathenau (the Hague), Staatsuitgeverij, 1980, 132 pp.

NEW ZEALAND

Computers, jobs and female employment

Women play an important and growing role in the workforce. From 1936 to 1976 the percentage of women actively engaged in the workforce rose from 24.9% to 36.6%. But what is going to be the effect of new technology?

Already concentrated in a narrow range of low-paid, low-level jobs subject to high unemployment, the New Zealand Public Service Association (PSA) sees computerisation making female workers even more vulnerable.

Even in the computing operations the best paid and highest status levels are dominated by men; low status occupations are held by women. The figures for the public service are representative:

Job	Males	Females	Females % of total
Programming	229	70	23.4
Terminal operator	1	489	99.8

And even the operator jobs are under threat because of the spread of devices such as "desk top" computers which allow users to have direct access to the main computers.

International forecasts on job losses for clerical workers vary. In drawing up an estimate for New Zealand, the PSA applied the reduction rate of 20% forecast by the European unions, of 30% by France's "Nora Report" (see S.L.B. 3/78, p. 207) and of 40% by Germany's Siemens report, to the 1976 clerical employment census figures.

These predictions of female job losses, says the PSA, deserve to be taken seriously by the New Zealand workforce. Not only would women's place be in the home - whether she likes it or not - but it will result in fierce competition between men and women for the remaining jobs, conjuring up recourse to a "fake" solution to the unemployment problem - replacing women by unemployed men.

The ramifications go far beyond the impact on female employment. While the answers vary, the fundamental question, says the PSA, remains "will the new technologies mean more jobs or less?" Any optimism must be measured against more than 50,000 people currently out of work. Moreover, the New Zealand Computer Society has predicted that by the year 2000 there would be 500,000 jobless.

Potential job losses for women clerical workers

Industry/occupation	Women employed 1976	No. if job levels reduced by 20%	30%	40%
Finance/insurance/ real estate/ business	34 904	27 933	24 433	20 942
Typists/machinists	33 866	27 093	23 706	20 319
Bookkeepers/cashiers	16 172	12 937	11 320	9 703
Telephone/telegraph operators	5 568	4 454	3 897	3 340
Other clerical	45 645	35 516	31 951	27 386
TOTALS	136 155	107 933	95 308	81 690
TOTAL JOB LOSSES		28 222	40 847	54 465

Manufacturing and assembly work is also being affected as robots and mechanical production take over the factory floor. Even supervisory jobs are affected because there is less to supervise.

Communications are also being transformed. The switch to subscriber dialling means that in the next 10 years the jobs of 3,000 operators will disappear. Other post office workers will be hit by computerised mail services (see also

S.L.B. 1/81, p. 10). The printing and retail trades are other areas where new technology has made fast inroads.

Over-all then, new technology presents a whole new set of problems in the workplace. Moreover, it is not possible to predict with any accuracy what the impact will be as the variables are both important and complex, including the development of the economy and socio-economic and industrial structures.

New Zealand unions have not ignored new technology and many unions have already negotiated job security, retraining and health clauses. But this is not enough. Unless there is a national commitment to job creation, the question of the income level of the unemployed will be one of the most difficult questions to solve in a microelectronic society.

An inter-union working party recommended a national trade union strategy to technological change based on the principles of consultation and negotiation (see also S.L.B. 4/80, p. 381); job security and a shorter working life; the protection of workers' health; and the right of all to a minimum income and to the opportunity to lead a creative and self-respecting life.

It also recommended that a small, full-time and paid research and training body be set up by the unions to monitor the constantly changing information on technology, to train union officials and to serve in an advisory capacity.

Source: PSA Research Discussion Paper No. 15: The new technology and the employment of women (Wellington), Jan. 1981, 18 pp.
Report of the inter-union working party on technology, 1979, 47 pp.
NZ Federation of Labour: "The Crunch". The effects of new technology in the workplace; the unions' response, 23 pp.

NORWAY

Government faces up to the 1980s

"The arrival of information technology raises a number of questions relating to the working environment and employment, which will be the focus of public debate in the coming years." This is the opinion of a special multidisciplinary working group set up by the Norwegian Department of Municipal and Labour Affairs.

The report is mainly based on existing literature and research but includes the results of new research carried out by members of the working group.

One of the features of the report is the way the authors have avoided turning it into a limited technical exposé and have instead endeavoured to deal with several important development factors within the context of the subject-matter. This means for example that, although the setting is Norwegian, international developments are constantly being brought in and integrated with the main theme.

Special attention is paid to information technology which is seen as the linchpin for technological development in the 1980s; "It is a basic technology with applications in nearly all areas within society and thereby an important source of change within society" says the report. It continues by soberly adding: "The spread of technology will to a great extent be governed by organisational economic and political factors which form part of the decision-making processes concerning use of new technologies. Developments should not necessarily be dramatically different from what has gone before. A sudden fall in employment caused by technological development is therefore not very likely".

The use of information technology appears to be affecting the employment situation of an increasing number of people both negatively and positively. The challenge lies in utilising technology to abolish monotonous and dangerous work, and to the greatest extent possible in overcoming the negative aspects before they have time to appear.

Changing job structures will become an increasingly important factor on the labour market. The "how" and "why" of the changes will therefore have to be thoroughly understood in order to prevent serious problems. In the opinion of the working group, the repercussions on the organisation of the labour market has a decisive influence on understanding the consequences of the new technology on employment and the working environment in the years to come.

The working group does not exclude the possibility of technological or "structural" unemployment in Norway in the 1980s but tends heavily towards the opinion that the country's general economic development, as well as internationally, will be the major determinant rather than technology itself.

According to the working group, an active industrial policy, including energy policy, seems to be the best guarantee for Norway's employment and working environment in the future decade. Not least from the point of view of employment, the public sector is considered crucial. It controls 35% of total employment and accounts for about 55% of the gross national product.

Three possible strategies are suggested for the 1980s - a growth strategy, a distribution strategy and a "wealth strategy". The growth strategy aims at long-term economic growth. The distribution strategy concentrates on employment which would mean that a substantial part of Norway's oil revenue would go into employment-creating measures both in the public and private sector. The last strategy is based on short-term measures to increase consumption by fiscal means and by using oil revenues. Although the working group does not make any recommendations it is evident that they consider the latter method too fraught with dangers from international economic developments.

Source: Norway: Department of Municipal and Labour Affairs: Sysselsetting og arbeidsmiljø i 80-arene, NOU. 1980:33 (Oslo), 93 pp.

SWEDEN

Computerisation: Government explores impact on employment

Two government commissions were set up in mid-1978, one to make a thorough investigation of the future effects of computerisation on employment, the other to specifically study the pace and pervasiveness of computerisation in industry and commerce. The broad aim is to consider the extent to which different courses of development have desirable or injurious effects on employment. Emphasising the rapid developments in microelectronics and the need to take these into account, the Minister of Labour, in proposing the establishment of these commissions, also mentioned the impact of industrial robot systems. About 600 industrial robots were installed in 1977 and are considered to have a large potential in engineering and other industries.

Source: Sweden: Ministry of Labour: Effects of computerisation on employment and working environment (Stockholm), Dir. 1978:76, p. 8.

Rationalisation measures threaten jobs in the telecommunications sector

A working party of the Telecommunication Section of the Swedish State Employee Union (SF) has studied the impact on jobs of planned rationalisation in the telecommunications services. It concluded that 7,700 jobs would be lost by 1991, more than half of these would disappear by 1986.

The rationalisation measures include:

(1) the introduction, in 1983-84, of an administrative computer system leading to job losses in the sales and planning department (30% reduction in sales staff according to the administration's estimates against a 50% decrease according to the union's and a further 20% decrease, or 320 jobs, in the planning department

in 1985-86);

(2) a new system of telephone billing which would entail 325 job losses;

(3) the replacement of manual mobile telephone exchanges by automatic ones in 1989-90 which would cause 500 job losses among telephonists;

(4) the introduction of electronic exchanges and maint tenance centres by 1991-92 would entail the loss of 1,400 technical jobs.

Earlier forecasts by the administration do not show any reduction in current employment levels. It explained that different figures as compared with the union's were due to the fact that its own assessments were based on an increase in traffic volume, they were calculated on shorter time periods and they did not cover all rationalisation projects.

The trade union's conclusions have prompted the administration to set up a joint working party to review the problem, to streamline the estimates of the employment impact of rationalisation and to establish guidelines for future manpower planning.

Source: PTTI News (Geneva), No. 7, July 1980, p. 5.

SWITZERLAND

Microprocessors and office work: government, employer and trade union opinions

While a recent study forecasts that the introduction of micro-electronics would entail the loss of some 250,000 office jobs in Switzerland by 1983, a government representative, as well as employers and trade unions, have reacted more optimistically.

On the government side an economic spokesman has pointed to the difficulties of forecasting the economic and social repercussions of new technology. Experience had shown that when labour demand for a particular job decreased, labour transferred to other sectors. However changes in price structure influence demand; a fall in real prices of certain goods resulting from the introduction of micro-electronics may increase demand for such goods since they are now cheaper but it also means that new needs can be met, providing the possibility of job creation and reduced working hours. Moreover, increased life expectancy brings with it increased demand in certain areas of the services sector, and a reduction in administrative costs made possible by electronics should also help to maintain jobs in the production sector.

A representative of the Central Union of Swiss Employers' Associations, while agreeing that new technology entailed job losses and involved adjustments, particularly in the areas of training and skills, warned against taking a simplistic or alarmist view of the matter. Until now the office equipment sector has been the smallest item of expenditure both for industrial undertakings and services (average investment of about 10,000 francs per workpost in offices as against 50,000 francs in the production sector; 1 US dollar = 1.65 Swiss francs). There have been very few basic technical innovations in the office equipment sector where productivity had increased very little (an average of 1% per year). Micro-electronic equipment does not necessarily need very large investment and it replaces equipment which did not require a large outlay. All these factors seem to indicate that data processing has a good chance of developing rapidly in the office equipment sector, where labour is already scarce even during the present recession.

According to the same source, it is forecast that jobs will be created in research laboratories, consulting engineering firms particularly those dealing with software, in services meeting new requirements and even in the production sector where a certain expansion is foreseen.

In Switzerland there are far more clerical vacancies than applicants and this situation will probably continue during the next few years (for there will be a considerable decrease in the number of young people coming into active life). The introduction of new technology, by developing the services sector and meeting the new requirements of its clientele, will therefore help to solve the problem.

He nevertheless concluded that the risk of technology-induced unemployment in offices should be taken seriously and that both sides of industry ought to keep a close watch on it. Indeed the Central Union of Swiss Employers' Associations, in conjunction with the Swiss Clerical Workers Association, has already set up a joint working group to discuss the matter.

Finally, a representative of the Swiss Federation of Trade Unions recently stated that although the unions were not against technological progress there was a need to protect the workers against any adverse consequences. "The coming technological upheavals and the resulting structural changes give a new meaning to union demands for workers' participation." As to the effects of micro-electronics on employment, the view was expressed that certain forecasts, in particular the one in question, were sometimes made lightly and did not take sufficient account of the positive effects that they may have on employment.

Source: "Micro-électronique et emploi: un certain espoir, W. Jucker (government spokesman on economic questions) s'élève contre les alarmistes", by F. Troxler, in Lutte Syndicale (Berne), No. 8, 20 Feb. 1980.
"Les nouvelles technologies dans le secteur des

bureaux" by H. Allenspach (representative of the Central Union of Swiss Employers' Associations, in Journal des Associations patronales (Zurich), No. 6, 7 Feb. 1980, pp. 113-117.
USS: Correspondence syndicale suisse (Berne), No. 43, 21 Nov. 1979.

"Talking typewriters" for the blind

The Swiss subsidiary of IBM, manufacturers of office equipment and computers, announced at the end of April 1980 the launching of audio-typing units to facilitate the employment of the blind and the visually impaired.

The unit transforms typewritten material into sounds, by means of a microprocessor. It produces synthetic speech with an unlimited vocabulary, enabling the typist to obtain all the information needed for the text pagination, text layout, margin, tabulation, punctuation and spelling (including use of capital letters). The typist "hears" what he types, he can "re-read" the text, ask the audio unit to spell words and receive additional instructions ("insert magnetic card", "end of ribbon", etc.). The playback speed can be controlled by the typist.

Synthetic speech is creating by combining the basic phonetic speech sounds stored in the unit's memory in accordance with pre-programmed pronunciation rules. The phonetic sounds are then generated by a voice synthesiser unit and transmitted through a headset or through a console speaker connected to the audio unit.

The audio unit may be used with magnetic card typewriters or with typewriters equipped with "memospheres" (already standard equipment in a number of firms).

French, German and English-language versions of the unit are obtainable (including one specially programmed for American pronunciation).

There are approximately 10,000 blind or visually impaired persons in Switzerland, one-third of whom are of working age. This new device will allow them access to office jobs from which they have been previously excluded because of their handicap.

Source: IBM Switzerland, Press Conference (Lausanne), 29 Apr. 1980.

UKRAINIAN SSR

Computer techniques training in secondary school

An experimental training programme in electronic data processing for secondary schools was drawn up recently in the Kiev Institute of Cybernetics.

Spread over three senior years, the programme combines theoretical and practical instruction in data-processing methods. Graduates from the school will receive, in addition to their diploma of general secondary education, a certificate enabling them to work as computer programmers and operators.

Source: Pravda Ukrainy (Kiev), 10 June 1979, p. 4.

UNITED KINGDOM

Government subsidises micro-processor training

The Department of Industry has approved proposals worth half-a-million pounds (1 US dollar = 0.48 pounds sterling) to promote short courses on micro-electronic systems for engineers in industry (see also S.L.B. 1/79, p. 7). The Micro-Processor Applications Training Scheme provides up to 50% of development costs, including equipment costs, and approved courses will be available in 13 regional training centres throughout the country. Plans to brief the country's 50,000 top "decision makers" on the potential applications of micro-processor technology, at a cost of 10 million pounds, are also being completed by the Department of Industry.

At the same time, the Department is supporting the development by the Open University of an awareness course for managers and a technical course for engineers. Managers will learn about the level of investment needed to develop systems using the new technology and the effects on company structure and staffing.

A further programme, costing 12 million pounds, will be developed by the Department of Education and Science to develop young people's understanding of micro-electronics.

Source: Trade and Industry (London), HMSO, 26 Jan. 1979, p. 158.
Department of Employment Gazette (London), HMSO, Vol. 87, Feb. 1979, p. 110.
Financial Times (London), 27 Feb. and 9 Mar. 1979, p. 11.

ASTMS reviews employment effects of new technology

A major discussion paper has recently been published by the Association of Scientific, Technical and Managerial Staffs (ASTMS). It gives a great deal of background information about the impact of microelectronics and the trend towards automation and unmanned factories.

The paper repeats the general approach to micro-electronics technology - it recognises potential benefits but fears that jobs will not be created as quickly as they are lost. It does argue, however, that high unemployment is inevitable, even if the correct remedial actions are taken. Listing other estimates, the paper forecasts the job displacement effect of the new information technology will range from between 10% to 30%, with certain specific areas suffering even higher impacts.

Taking the official government figures for labour supply in 1985 and 1991 - 25,538,000 and 26,169,000 respectively - the assumption is made: that by 1985 the number of information workers in the economy will reach 50% and will remain constant through 1991; that job displacement due to new technology on information workers will be 20% in 1985 and 30% by 1991; and that the unemployment rate among all other workers will be a static 10%. The consequences of these assumptions are an unemployment rate of 15% in 1985 and of 20% by 1991; i.e. 3,831,000 and 5,235,000 people, respectively.

Accepting the argument that results depend upon assumptions, the ASTMS feels that their own assumptions may be all too plausible: there is no escaping the magnitude of the potential unemployment problem. Even forecasts which predict high growth rates with higher productivity show levels of unemployment varying from 11.9% in 1991 (by the Institute of Manpower Studies) to around 16% in 1990 (by the Cambridge Economic Policy Group).

The benefits of new technology will accrue at national level with higher gross domestic product, a stronger balance of payments and higher productivity, but the costs will come in higher unemployment. The ASTMS therefore feels that much more work needs to be done on assessing the impact of technological change. Trade unions must be informed about the options and have increased influence over the way technology is used. The main preoccupation is how to protect jobs. Methods of doing this include reduced overtime, shorter working weeks, more holidays, sabbaticals, and worksharing. On top of this flexible retirement age and paid educational leave will help to ease the situation.

All this implies a change of attitude towards work. The question of income distribution must be separated from that of employment. There is a need to construct a new system of income distribution which will maintain people's standard of living when not in employment.

Source: ASTMS: Technological change and collective bargaining. A discussion paper (London), 1979, 56 pp.

Job displacement and new technology: a TUC response

Recognising that developments in electronics are creating a late twentieth century industrial revolution, the Trades Union Congress (TUC), in an interim report on new technology, urges trade unions to conclude "new technology agreements" so as to combine the development of collective bargaining machinery with a participative and flexible response to technological developments. The report is part of the over-all preparation of a union strategy to be finalised at the annual TUC Congress in September 1979.

The proposed agreements would involve the unions in looking for opportunities for introducing new processes and products on the basis of agreed plans. Joint management/union teams should be set up in every enterprise to assess the implementation and monitoring of technological change. The aim should be to come up with plans tailored to fit individual enterprises.

Joint teams should have full access to information on manpower and turnover levels and plans should cover product development, output, investment and employment. The TUC commends the recent agreement between the Confederation of Shipbuilding and Engineering Unions (CSEU) and Lucas Aerospace (see S.L.B. 2/79, p. 171) which shifted the company from a policy of plant closures to one of developing alternative products.

The first principle of the new approach would be that no new technology which has major effects on the workforce should be introduced unilaterally. Other conditions for the smooth introduction of technological changes would include: (1) wherever possible, a guarantee of job security for the existing workforce; (2) guarantees to workers, whose jobs are reorganised, that their individual earnings and status will be maintained; (3) priority to be given to the existing workforce in retraining for new skills; (4) adequate redundancy provisions, preferably in the form of continuing payments rather than lump-sum payments to workers.

At every stage, says the report, technological change should be linked with a reduction in the working week, working year and working life-time. Priority should be given to moving towards: (1) the 35-hour week; (2) a reduction in systematic overtime; (3) longer holidays; (4) better provision for time off for public and trade union duties; (5) sabbatical leave; and (6) early retirement on improved pensions.

These measures should be seen not just in the context of sharing out less work to avoid increasing unemployment, but in the more positive light of setting job creation targets for enterprises and accompanying them with measures which enable workers to have some of the benefits in the form of increased leisure.

Moreover, new technology should not be used to deskill or fragment jobs, or introduce greater routine. This is harmful to both job satisfaction and productivity. The responsibility and autonomy of work groups should be increased, job content enlarged and single status introduced for all employees. Negotiators must try to ensure that technological change does not produce inequitable differences within a workforce - either in employment prospects, job content or pay.

But union involvement, says the TUC, in the introduction of new technology at plant or company level is not enough. At national level, the right economic, industrial, social and manpower policies must be pursued - particularly measures to promote industrial research and development. Already, due to productivity increases and the 200,000 or so new entrants to the labour force each year, economic growth rates in excess of 3% will be needed to reduce unemployment. An over-all economic and social plan was needed and this meant co-operation between the Government, the corporations and the workforce.

Finally, the TUC points to the need for a more detailed study on the impact of technological change in practice. A practical solution to the problems would have to be based on more fact finding.

Source: TUC: Employment and Technology: A TUC Interim Report (London), 1979, 50 pp.

Microelectronics and job displacement: a white-collar union analysis

The Association of Professional, Executive, Clerical and Computer Staff (APEX) was one of the first trade unions to set up a working party to analyse the effects of microelectronics in a sector of industry where it is anticipated far-reaching changes will take place (see also S.L.B. 1/79, p. 7, 2/79, p. 128).

The working party's first report, subsequently endorsed by the APEX Executive Council, sets out the background to the current technology debate. It makes clear that if the effect of micro-electronic technology is only to prevent the number of white-collar jobs from continuing to increase, this alone will mean greater unemployment. It points out that women's employment is mainly in the office sector and argues that women's incomes are a crucial factor in keeping low-paid families out of poverty, in view of the fact that around 30% of working women with children are the sole family income earner.

The report also refers to the large number of young people who will be coming onto the labour market over the next five years, pointing out that an extra 1.2 million new jobs will be required by 1982 in order to keep unemployment at its present level.

Rejecting the view that the productivity increases

which new technology makes possible should be opposed by trade unions, the report argues that jobs could be "as much, or more at risk" if new technology is resisted or ignored. Unions should co-operate with employers and the Government to prevent what could be large-scale unemployment among office workers over the next decade.

The report lays down a detailed negotiating strategy for union representatives to follow at company level in dealing with employer proposals to introduce word processing and similar advanced office technology. The main steps are to find out what the employer is planning, and insist on joint discussion; to find out what the motive of the employer is for introducing the system; to make detailed assessments of the likely effects on manpower; to begin negotiations on shorter working hours; to negotiate where necessary on diversifying the company's activities into new and expanding areas; and to ask for adequate assurances on retraining, redeployment, etc.

APEX is against "natural wastage" as a solution, which it says simply redistributes unemployment to the potential jobseeker just out of school.

Whatever happens to the number of jobs there will be profound changes in their content as a result of computerisation - it can either be used to dehumanise and mechanise jobs, or to improve their quality. The report moreover concludes that while typists and clerks will be the first to be affected, even very senior managerial jobs are likely to change drastically in the longer term. Job evaluation studies are needed to prepare for that change.

The report also tackles the concrete worries of trade union members of the hazards of working with computer terminals, and in particular visual display units (VDUs). It sets out minimal acceptable standards for technical details of VDUs, including screen colour, office illumination, character size, flicker rate and screen glare (for more details on eye fatigue of VDU operators see S.L.B. 3/78 p. 277).

The report concludes with a number of key recommendations. A major programme of union education and training is proposed, it is also suggested that in each workplace a senior staff representative should be given specific responsibility for issues related to computerisation. More efforts need to be made to encourage joint negotiations with other unions on this issue. And company, rather than plant level, agreements should be sought wherever possible. The working party expresses the view that tackling new technology at company level is an excellent way of bringing real industrial democracy closer. Reference is also made to the need to keep in close contact with unions in other countries on their strategies for new technology.

APEX is already following the policies outlined in the working party's report.

Source: APEX: Office technology: The trade union response (London), 1979, 68 pp.

Employment impact of new technology

The Manpower Study Group on Micro-electronics established by the Department of Employment in July 1978 to examine the manpower implications of micro-electronic technology up to 1990 has recently published its report. It relies heavily on case studies conducted in Japan and the United Kingdom.

From the beginning, the authors of the report assume that there is no other option but to adapt to micro-electronic technology and, at the risk of sounding complacent, caution against too dramatic a view of technological change. Tracing the economic background, the Study Group makes the following points: (1) while the over-all employment effect is virtually impossible to gauge, past experience suggests that, in the long run, technological change has been beneficial to both output and employment; (2) the pace of technological change is more likely to be gradual and persistent rather than sudden and revolutionary; (3) some evidence suggests that increased productivity may release manpower from industry to the tertiary sector. However, the fact that secondary and service employment tend to grow together, though at different rates, suggests that the growth of employment in services is not necessarily associated in the long run with a loss of jobs in the secondary sector; (4) many of the models produced in relation to micro-electronic technology are unlikely to help in forecasting the future as they often ignore the complex interplay of macro-economic and technological factors.

The importance for British industry of keeping up with its competitors in its use of automated assembly techniques including robotics, is stressed. The evidence, says the report, is that industry is being slow to seize these opportunities.

Micro-electronic technology will introduce two kinds of process changes: some will enable production to be carried out on a more continuous basis; others will substitute machines for men without necessarily changing the production process. In the next ten years the former is likely to have a greater effect upon employment than the latter. But, says the report, the over-all effect will depend crucially upon whether new technology is used as a means of expanding output or simply as a means of cutting labour and costs.

In areas which are already highly automated, the main effect will be the introduction of more sophisticated control systems, with negligible effects on over-all manpower levels.

Undoubtedly there will be people displaced, but the Study Group also sees new opportunities for market growth and hence for employment. Moreover, there are economic and behavioural reasons for suspecting that the new "electronic office" will not easily be translated into reality. Experience with microprocessors suggests that productivity gains and resulting job displacement are not necessarily the automatic result.

Loss of job opportunities are, however, foreseen in the less skilled clerical and subclerical areas in the service sector, in the telecommunications manufacturing industry and for those school-leavers who have low educational qualifications.

According to the report the successful introduction of innovation relies heavily on labour-management consultation. That large companies in the United States and Japan had gone ahead with introducing micro-electronic technology fairly rapidly was partly due to such consultation ensuring no compulsory redundancies. Recent union proposals for new technology agreements offer a constructive approach (see S.L.B. 3/79, p. 230). However, says the report, for such proposals to be acceptable, trade unions will have to offer a greater degree of flexibility as a return for the employment guarantees they seek. Moreover, in general, the organisational impact of new technology is likely to reinforce existing arguments against work sharing as a means of combating unemployment (see S.L.B. 2/79, p. 126).

Contrary to some views, the Study Group felt that the pace and pattern of micro-electronics does not so far justify a radical restructuring of existing government policies in the employment field. It opted rather for setting up "early warning signals" and recommended that the Department of Employment should:

(i) monitor developments in certain manufacturing industries - especially those using computer aided systems;

(ii) give more attention to developments towards the convergence of particular technologies which may have machine/man substitution effects, e.g. integrated use of machine tools, robotics and materials handling systems;

(iii) monitor redundancies due to new technology and the effects of "no redundancy" agreements in connection with investment in new technology;

(iv) continue studies to relate the manpower needs of new technology and the requisite training;

(v) give special attention to monitoring the effects of introducing new office technology;

(vi) promote experiments within the civil service using electronic office systems;

(vii) encourage training organisations to relate their efforts more closely to the skill requirements of micro electronic technology; and to pay special attention to the problems of the relatively unskilled.

Source: United Kingdom: The Manpower Implications of Micro-Electronic Technology (London), HMSO, 1979, 110 pp.

Will microprocessors wipe out secretarial jobs?

According to a recent report by "Youthaid", a pressure group concerned with youth unemployment, the probable number of word processors in the United Kingdom is 9,000. Estimates of annual market growth from various manufacturers and consultants range from a cautious 15% to around 40%.

The fairly wide range of uncertainty leads the author to talk in terms of scenario building rather than forecasting. However, even taking the lower figure, the scenario shows a total job loss in excess of 36,000 over the next 10 years. Taking the higher growth rate of 40%, then the number of typing jobs lost by 1989 will be over 260,000.

While the report looks at several possibilities recouping this loss, it tends to discard them and cites 10 case studies all of which show staff reductions after the introduction of word processors. Moreover, although micro-technology will create new jobs it is extremely unlikely such job creation will be on a matching basis. And, because most jobs involving typing are carried out by women - it is women who will be hit the hardest. The problem will be aggravated by the fact that the female labour force is expected to increase both in real terms and as a proportion of the total labour force.

However, there is some hope in the fact that the process is not going to take place overnight. The author grants a breathing space of 10 to 20 years - during which time society must decide whether it will be a blessing, freeing many people from boring work, or on the contrary, consigning them to idleness and poverty.

Source: David Taylor, Cheap Words ...?, Youthaid (London), Oct. 1979, 21 pp.

Layoffs spark claim for shorter week in computer company

One of the big companies involved in the design, manufacture and marketing of electronic systems is International Computers Limited (ICL). However, according to unions at ICL, for a company with a vested interest in change and new systems of work, ICL has shown reluctance to give real consideration to shorter working hours for its own employees.

A claim for shorter working hours within the company is now being led by the Association of Scientific, Technical and Managerial Staffs (ASTMS). In a specially prepared document the ASTMS describes the computer industry as fast-changing and, as a result, full of stress, with fears about job security and skill obsolescence dogging its employees.

Going on past evidence, says the document, technological changes have led to a reduced demand for labour. In 1971 the UK computer industry employed 53,000 people. However, by 1977 this had fallen to 43,000, a 19% reduction in 6 years.

The principal cause, according to the report, was the switch to smaller, integrated circuits which are suitable for more capital-intensive assembly methods.

Technologies are still changing rapidly. The continuing development of integrated circuit memories, for example, will affect computer manufacture, while more automated methods of production are being developed. In addition, the industry is its own customer. ICL, for example, is one of the first users of its own new word-processing system.

For all these reasons, plus the 1,000 layoffs caused by closing one of ICL's plants earlier in the year, the union is now prepared to take issue over a call for working less hours in the week.

The document also deals with companies looking for greater productivity and high profits - like ICL. It says: "Such co-operation is unlikely to be forthcoming if the reward is more unemployment, increased job insecurity and more stress".

A shorter working week, says ASTMS, will play a major part in overcoming these problems. It will be one of the main ways in which the benefits of industrial change will be distributed to employees.

Source: ASTMS: The case for the shorter working week in ICL - the claim submitted by the Joint Staff Trade Unions, 11 pp.

New technology, employment and skills: some generalisations questioned

A 1980 paper published by Youthaid, an independent research group, describes recent trends in employment levels and skill structures in four key industries: machine tools, electronic components, iron castings and retail distribution. This paper examines some of the changes in these industries over the past decade, and uses them to illustrate a number of points.

The first is that although impressive, the silicon chip is only one facet of industries continuing technical development. But how many people know of the dramatic increase in productivity brought about by the replacement of traditional cutting surfaces by diamonds in the machine tools industry, or are acquainted with the impact that chemical binders or continuous sand mixing machinery have had on the foundry industry?

Secondly, the paper examines the widespread assumption that new technology will tend to displace traditional craft skills while creating employment for semi-skilled machine operators and highly trained technicians and technologists. This might be so, but the electronic components industry - no stranger to new technology - employed a higher proportion of skilled craftsmen to produce a given unit of output in 1977 than was the case in 1970, while it employed a lower proportion of technicians. During the same period, the proportion of scientists and technologists in the machine tools industry also fell, while the proportion of technicians actually increased.

Thirdly, and in view of recent debate perhaps most important, the paper shows that if retail distribution is typical of the service sector as a whole, this sector cannot be relied upon to mop up labour displaced from manufacturing. Retail distribution is now less labour intensive at all stages than it has ever been before, and the trend seems certain to continue. Supermarkets are less labour intensive than the grocers they replaced, and this pattern is repeated in all branches of the non-food sector, from records to refrigerators.

Source: Youthaid: Employment and occupation structure in four industries, by David Taylor (London), 1980, 24 pp.

Adopting new technology: experience in engineering and electronics

According to a recent report by "Youthaid", a pressure group concerned with youth unemployment, sweeping generalisations about the likely employment impact of technological change in manufacturing industry can often be misleading. In some cases, it might lead to cost-price reductions that stimulate demand to such an extent that increased employment levels result. In other cases the labour-saving effect will not be offset and the result will be job losses.

The report itself was the result of a 2-year study into the effect of industrial change and capital investment on employment prospects in the iron castings, metalworking machine tools, and electronics components industries. The over-all survey results are based on replies representing about 35%, 42% and 47% respectively of establishments in each industry with 20 or more employees.

Labour-saving investment, says the report, does not automatically result in a loss of employment opportunities in an industry. The extent to which industry can produce saleable products is probably of infinitely greater importance in determining over-all employment levels than the way in which it manufactures them. This can be seen by looking at recent developments in iron castings and electronics. In the first case modest improvements in per capita productivity were accompanied by a precipitous fall in employment levels, while in the second case, big increases in per capita productivity have been accompanied by rising employment levels.

New office technology will probably have the effect of reducing the number of management, administrative and clerical staff required by many firms, though this could be reversed if sophisticated manufacturing systems result in increased administrative effort.

The report sees a continued fall in the demand for unskilled manual workers but notes little evidence to support the popular belief that new production technology is going to bring about a general shift into semi-skilled manual work. Proponents of this belief overlook the case with which simple machine-minding jobs can themselves be either automated or brought under the control of a single operator- perhaps a technician or skilled worker controlling a large number of machines from a single console.

The impact of technology on skilled manual workers is less clear. In spite of the skill displacing effect of much new technology, it is by no means certain that there will be an over-all decline in skilled manual work. Although this will happen in some industries (certainly in iron castings), new technology will increase the need for flexibility in the use of manpower. This could involve a blurring of current demarcation lines between technicians and skilled workers - just as employment trends over the past two years among manual workers in the machine tools and iron casting industries reflect the deepening economic recession, employment among technologists, scientists, technicians and draughtsmen in the electronic components industry seems to have been growing in step with technological progress in the industry.

Finally, the report briefly examines the ability of one of the major industries in the service sector - retail distribution - to provide employment for workers shed by manufacturing industry. It sees no possibility for this sector to soak up the numbers of people likely to be leaving manufacturing. On the contrary, this sector itself has terrific potential for introducing labour-saving innovations, so that its own manpower requirements will be reduced. Even a considerable demand in this area will not necessarily generate an appreciable increase in employment.

Source: Youthaid: Innovation and Employment, by David Taylor (London), Feb. 1981, 120 pp.

When 8 million are jobless

Another voice has joined the series of dire warnings on the consequences of microchip technology. Well-known for its consultancy services, Ashridge Management College (AMC) has produced a report which shows that 60% of chief executives of large companies anticipate radical change - triggered off by the electronic office and falling microelectronic costs.

At the same time, AMC envisages that the next 18 years will see unemployment rising to 8,000,000 in the United Kingdom.

The report also suggests that increased competition and increased working capital requirements could lead to the business failure of many smaller firms and the formation by merger and acquisition of substantially larger units. Specialist companies, whatever the size, will survive.

Source: Computer Weekly (London), 12 Feb. 1981.

Microelectronics in education

The Government has announced a scheme which aims to put at least one microcomputer into every secondary school by the end of 1982.

The scheme, which went into operation on 1 June 1981, is part of the wider attention the Government was giving to information technology.

Source: British Information Service: Survey of Current Affairs, (London), Apr. 1981.

UNITED STATES

Postal workers gain job security through mediation

Under a national contract covering 1978-81 more than half a million postal workers achieved lifetime job security and more effective cost-of-living protection. Because of mechanisation, job security was a crucial issue in the negotiations.

The final terms were set by a special mediator, after complete deadlock in negotiations threatened to lead to industrial action.

The main unions involved were the American Postal Workers' Union (299,000 members) and the National Association of Letter Carriers (200,000 members).

Job security. In the dispute, the postal management had tried to weaken the "no layoff" clause first negotiated in 1971 for all employees (regardless of seniority), and a major union priority. The mediator, however, ruled that all those in the regular workforce as of 15 September 1978 are "protected atainst layoff" for the rest of their working lives. New employees would receive such lifetime job guarantees after six years of service, and the employer cannot lay off an employee solely to prevent him reaching permanent status.

Preconditions to layoff. Any layoff must be preceded by 90 days' advance notice to the union and 60 days' advance notice to the employee concerned. The employer's advance notice must contain the reasons for the proposed layoff.

Prior to any layoffs, the employer must first terminate casual labour, minimise overtime, decrease part-time work, solicit early retirees with the inducement of a lump-sum severance pay, and voluntary reassignments to existing vacancies. Where a job is abolished, and the employee cannot be placed in the same grade, his rate of pay is protected until such a time as he fails to apply for a position in his former wage level.

Layoff procedures. Layoffs will be made according to agreed, but fairly complicated, procedures which include seniority criteria. Employees who are laid off shall be placed on recall lists for a period of two years.

Technological and mechanisation changes committee. This joint management/union committee operates at the national level and its job is to try and resolve any questions resulting from new technology. The union committee members must be informed as much in advance as possible of technical or mechanisation changes which will affect jobs, wages, working conditions or create new jobs. 90 days' advance notice must be given before new machinery is installed.

Under the agreement, any new jobs created by technological or mechanisation changes shall be offered to present employees capable of being trained to carry out the new functions, with suitable training provided by the employer.

Concerned about the dehumanisation brought about by the monotony and tedium of mechanised jobs, the unions have also negotiated a proviso in the new contract which gives them access to final data sheets used by the employer to establish work or time standards.

Wages. A major point won by the unions was the decision to lift the ceiling on annual cost-of-living adjustments and to grant annual increments linked to cost-of-living gains. Actual wage increases were moderate.

A National Joint Committee on Maximisation was also established in an attempt to maximise the number of full-time employees in each installation.

Source: US Postal Service National Agreement, 21 July 1978 - 20 July 1981, 80 pp.
APWU-AFL-CIO: The American Postal Worker, Mar. 1979.
PTTI: Conditions of work in the United States Postal Service by P.C. Bowyer (Geneva), Spring

1979, 59 pp.
NALC-AFL-CIO: The Postal Record (Washington, D.C.), Mar. 1979.

Unions face up to technology's impact on jobs

A call for a national strategy to counteract job losses caused by new technology emerged from a three-day conference on technological change sponsored by the AFL-CIO Department for Professional Workers in June 1979.

The 150 participants heard prominent experts from labour, government and industry put forward views on how the problems engendered by technology may be solved or modified.

Predicting more worker displacement by machines in future, the Chairman of the National Commission for Employment Policy warned that the days when workers didn't have to worry very much about job losses, due to expansion of the economy, were now over. Visions of a "technological explosion" to come were described by a panel of computer specialists, engineers and company representatives.

Union representatives described the problem of lost and downgraded jobs particularly in the highly-automated telecommunications industry and on the railways where automation had brought about a 65% drop in employment since 1944. Protection negotiated by the Railway and Airline Clerks included guaranteed negotiation before technological changes are introduced, provisions for natural attrition in layoffs, early retirement and maintenance of fringe benefits for displaced workers.

The AFL-CIO economist urged more federal regulations over the export of technology, claiming that such practice further exacerbates unemployment at home.

The Secretary of Labor for Employment and Training called for a partnership with the unions in creating "a comprehensive policy approach to move toward a situation where all workers who are displaced from their jobs involuntarily will be afforded the possibility of being retrained and, where necessary, relocated to new employment".

Source: AFL-CIO News (Washington, D.C.), No. 25, 23 June 1979.

Farm workers score point in job fight with machines

Farm workers whose jobs are threatened by mechanical machines have made some progress in slowing down the development of mechanisation.

A California Superior Court judge ruled in February 1980 that the University of California must participate in a trial to defend charges that it has unlawfully used federal funds to subsidise the development of labour-saving agricultural machines that benefit a limited number of private agricultural businesses, rather than the public at large.

A federally-financed legal aid group, California Rural Legal Assistance, brought the suit against the university, claiming the research was tailored to the needs of specific private companies. The court ruling followed a speech by the Secretary of Agriculture in which he strengthened a month-old policy statement declaring that as a general rule the Agriculture Department would not finance agricultural research whose major impact would be "the replacing of an adequate and willing workforce by machines".

The University is considered to be one of the most successful developers of mechanical harvesters. It is now extending the technology to develop machines to harvest lettuce and wine grapes. Californian farmworkers claim that by 1982 as many as 40,000 jobs may have been lost to the machines.

Source: The New York Times, 11 Feb. 1980.
The Economist (London), 16 Feb. 1980, p. 27.

A look at coming technologies: some employment implications

Speaking at the second World Computing Services Industry Congress, Alvin Toffler, author of "The Future Shock" and "The Third Wave", forecast a new social scenario as a result of computer-related developments. In this scenario, increased integration of video and electronic technologies will soon allow a growing number of executives and clerical workers to do their jobs without ever leaving their homes. Such a transfer of job functions from office to home will turn many homes into what Toffler calls "electronic cottages".

For individuals who will be unable to adapt to the growing occupational demands of the computer age, the author sees a need for a technological breakthrough in natural computer languages. This would make computer systems available for the first time to a large class of unskilled users who would otherwise find the systems forever inaccessible.

At the same time, James Martin, an authority on emerging data processing and telecommunications technologies, sees the use of robots expanding by 60% a year through the 1980s (see also S.L.B. 1/79, p. 9); factories being built in space for manufacturing semiconductors in pure, vacuum environments before 1990; semiconductors with as many as 10 million circuits on a silicon chip; optical fibres completely replacing copper wires in telecommunications; and video discs opening up a whole new storage technology.

Panelists at the National Computing Conference held at the end of May 1980 saw unrestrained technological development as a threat both to employment and to workers' identity. An adviser to the United Autoworkers Union cited robotics as posing a particularly dangerous threat to workers, with one out of every five jobs on automobile assembly lines being replaced by robots by 1985.

Automated offices were catching up on white-collar workers and would subject them to the same constraints as manual workers - rigid guidelines in areas such as output speed and error rate. Women will be particularly affected as 90% of such jobs are filled by women. People will be forced to make major behavioural changes.

These problems do not necessarily mean that the users will reject technology. They do suggest that the move into the "office of the future" will be costly both in human currency as well as in capital investment.

Source: Computerworld (Framingham, Mass.), 5 May, 2 June and 7 July 1980.

General Motors: sweeping switch to robotics

By 1983, the General Motors Corporation will have stepped up the number of robots used on its assembly lines from 150 to 800. By 1990, this figure will have risen to 13,000.

The new method replaces human welders with robots which assemble an entire car body in two or three stages. Normally, the process is done in 40 to 50 steps by workers using hand-held welding guns.

The company sees its new programme as only the start of a much more sweeping switch to robotics.

The United Auto Workers Union (UAW) estimates that assemly-line labour could be cut by as much as 50% over the next nine years because of robots and other automation projects. But, says the UAW, there is little it can do to stop the process.

Source: Business Week (New York), 16 Mar. 1981, p. 27.

INTERNATIONAL

Performing artists unions claim new technology erodes jobs

Representatives of performers' trade unions from 26 countries expressed concern over the impact of new recording techniques on job opportunities at a Symposium organised in Geneva from 10 to 12 January 1979 by the International Federation of Musicians (FIM) and the International Federation of Actors (FIA) (see also article on the FISTAV Congress in S.L.B. 4/78, p. 350). Participants examined the replies to a questionnaire sent to 30 member associations or unions.

Repeated use of recorded performances, together with direct diffusion to vast audiences simultaneously, have resulted in unprecedented technological unemployment which can only increase, as long as the performers themselves have no legal power to control the utilisation of their own work. According to the unions, while the work of performers becomes increasingly accessible internationally, the economic position of the performer has never been worse. If this continues it will eventually restrict cultural life in many countries.

The only international Convention recognising specific rights of performers is the Rome Convention of 1961. The Convention is theoretically applicable only to international situations, but obviously no government would be likely to grant foreign performers protection which they do not grant to their own nationals, and the Convention has only a handful of adherents.

However, in a declaration adopted by the Symposium, the performers' unions called for more effective control, made possible by exclusive rights in national legislation, over the utilisation of performances.

A related issue was the question of collection and distribution of remuneration arising from the broadcasting or communication to the public of phonograms published for commercial purposes (e.g. fee for the use of records). The Rome Convention, in Article 12, does provide for such remuneration but leaves open the question of the manner in which it should be distributed. The Symposium discussed a number of principles on how the collection and distribution of these monies should be undertaken across national frontiers. In general, most of the performers' unions agreed that the procedures for distributing the remuneration should be determined in the country where the money is due and not in the country where the money is collected, or in other words not where the secondary performance took place.

The Symposium also noted that in some circumstances, for example, when recordings are used in developing countries, bilateral agreements between collecting and distributing societies can be made so that the money can remain in the country where it is collected, that is in the country where the secondary performance took place. This principle recognises that the use of imported recordings has potentially harmful effects on employment and revenue of performers in the importing country, and that therefore some means to compensate national performers should be found.

Finally, the Symposium noted that the remuneration could be used for collective purposes (such as special funds or compensation for lost employment) or for individual distribution.

Source: FIM/FIA: Symposium on International Protection of Performers and Performers' Rights (Geneva), 10-12 Jan. 1979 (Background papers: ILO file - IC/12-0-4-1003-06-4).

Microelectronics: uncertainty over employment effects

The impact of microelectronics continues to be the subject of wide debate. While no one really seems to know whether the net effect on employment will be negative or positive, there is a tacit consensus that there will be a massive redeployment of workers with major social consequences.

While electronic equipment could put an end to 250,000 office jobs by 1983, according to research by the Association of Professional Executive, Clerical and Computer Staff (APEX) in the United Kingdom, the US consultant, Arthur D. Little, points to the creation of 1 million new jobs among producers - 60% of these in the United States. This claim is based on a 3-year study of markets in France, the Federal Republic of Germany, the United Kingdom and the United States, which forecasts that the use of microprocessors in home appliances in Western Europe alone will increase from about 1.3 million in 1977 to 128 million by 1987. The US market for numerical control systems, robots and automatic material-handling systems is expected to reach 16,800 million dollars in 1987.

However, recent data from the Bureau of Labor Statistics (BLS) in the United States indicate that academic institutions may glut the market for computer-related jobs by 1985. BLS projects that, from 1976 to 1985, average annual openings for computer programmers will total 9,700; for systems analysts, 7,600. These estimates are substantially lower than earlier BLS projects - 23,000 a year for programmers and 27,000 a year for systems analysts.

Meanwhile, the EEC Commissioner for Industry has called for a concerted European effort in the new technologies of electronics and communications. Action should include an integrated information network linking Community institutions and governments, a European satellite programme, an intense education and training effort and support for the European electronic component industry. If the programme was to be successful, people would have to be convinced that innovation did not necessarily mean unemployment.

Source: Financial Times (London), 1 Feb., 20 and 23 Mar. 1979.
Computer World (Boston), 9 Apr. 1979, pp. 1 and 8.

FIET comes to terms with computers and jobs

Following an international conference held in November 1978 (see S.L.B. 1/79, p. 42), the International Federation of Commercial, Clerical and Technical Employees (FIET) has drawn up, for its 19th World Congress (Caracas, 26-30 November 1979) an action programme to deal with the effects of computer technology on the jobs of workers in the sectors it covers. The Congress approved the action programme.

While accepting that trade unions need not automatically be hostile to computerisation, the FIET document argues that: (1) productivity gains from using micro-electronic computer equipment are much greater than ever before; (2) the world economy is not capable of growing fast enough to take up the growth in productivity; (3) worldwide unemployment is already intolerably high; (4) more people will enter the employment market in the next 5 years. As a result there is a grave danger that the uncontrolled use of computerised techniques in offices could result in large-scale job losses.

The FIET programme therefore lays down a detailed negotiating strategy which will represent a common approach on new technology from FIET's 179 affiliated unions in 80 countries, altogether representing 6.5 million members. The programme has six main components:

(1) Joint management-union manpower planning to assess the detailed manpower effects of alternative changes in technology.

(2) Collective agreements should link increase in productivity from new technology to expanded output and employment. Joint management-union committees should examine how technology can increase sales.

(3) Where the market is saturated, joint agreement should be reached on alternative goods or services.

(4) Unions should press for a reduction in over-all working life, including the 35-hour week, early retirement and 5 weeks' holiday.

(5) Trade unions should reject redundancy resulting from new technology. However, this is not sufficient to protect jobs. Agreements should therefore be based on future employment levels and policies based on non-replacement of workers leaving the firm should be rejected.

(6) Where redundancies are unavoidable, there should be advance notice, adequate retraining and help to find alternative jobs with employer-funded income maintenance during the interim period.

The programme also stresses the importance of job satisfaction, trade union participation in systems design, and safety and health.

FIET recommends that trade union action on new technology should be carefully co-ordinated at all levels: plant, national and international.

Source: Computers and work: FIET Action Programme (Geneva), Oct. 1979, 20 pp, and information provided by FIET.

OECD looks at links between the economy, innovation and employment

Social acceptance of a higher rate of technical change will depend on a satisfactory balance between the generation of new employment and the loss of old jobs, says the group of experts making up the OECD's Committee for Scientific and Technological Policy. The outcome of the two years' research, the Committee's work was based on four sectoral studies (electronics, machine tools, pharmaceuticals and fertilisers/pesticides) designed to identify: (1) "the impact of the past decade's economic and social changes on research and innovation" and (2) "the circumstances in which such activities can help our economies overcome the difficulties they are up against".

Investigating the links between the economy and the research-innovation system, which are necessarily of a long-term nature, inevitably leads to an examination of structural problems. These are summed up as slow economic growth, high levels of unemployment, prevailing inflation, the oil crisis, a new distribution of industrial power between countries and changed social values and aspirations.

Against this background, says the Committee, there has been a general slackening in innovation with research oriented towards short-term, low-risk projects. Electronics is a spectacular exception.

Accepting that the rate and direction of technical change affect general economic conditions, the Committee thought it more vital than ever before, to harness science and technology to economic and social policies. Examining three variables - productivity, prices and employment - revealed a disquieting trend. Since 1973, lower productivity, declining employment and rising inflation had been the rule. Only in the service sector has employment increased. Particular concern was expressed about the inability of economists to analyse the fundamental mechanisms of causation.

The main questions that needed an answer were: To what extent will the growth of the service sector compensate for the manpower reduction in manufacturing? Under what conditions will technical progress modify the nature of work and leisure by creating occupations increasingly remote from traditional production tasks?

There was need for greater emphasis on policies to stimulate economic growth. This need derived from the substantial underutilisation of resources, including high levels of unemployment, concurrent low rates of investment, with their detrimental impact on the growth of productive capacity.

However, if the service sector is to be the source of job creation, technical change must proceed at a pace and in directions which will ensure that the new activities offset job displacement. Hence the importance to governments of being aware of the potential problems involved in a strongly capital-intensive technical advance rather than a labour-intensive one. This is why the implications of technical change are at the heart of economic policy options.

Taking the view that governments have an active role to play in promoting innovation and that both the opportunities provided and the constraints imposed by science and technology must be taken into account, the Committee emphasises the importance of three objectives.

(1) Maintaining and improving innovative capacity within a long-term perspective. This means that basic research must be shielded from the consequence of recession.

(2) Sustaining a higher rate of technical advance and productivity increase. This means research into fundamental technologies - sometimes considered too risky by industry or too applied on the part of universities. Technological pluralism was needed to avoid being caught short, as in the energy crisis, by political or "technological" surprises.

(3) Promoting social innovation and technologies. This calls for special government support as demand is far less clear than in the marketing of consumer goods, and includes the quality of working life, collective services, working conditions and the educational and cultural framework.

According to the Committee, "the most intractable problems lie not in the potential of science and technology as such but rather in the capacity of our economic systems to make satisfactory use of this potential". Although the public has become accustomed to the economic management of society there is a long way to go before the full implications of managing technology are realised. Research and innovation must be associated more closely with other aspects of public policy - in particular economic and social policy.

Source: The OECD Observer (Paris), No. 104, May 1980, pp. 16-22.

Forecasting the impact of technological change on productivity and employment

A seminar on the impact of technological change was organised at the Centre for Industrial Studies in Geneva

(Switzerland) between 3 and 5 September 1980. Under the chairmanship of Wassili Leontief, Nobel Prize winner in economics, the meeting, which was attended by some 30 economists, planners and engineers dealt mainly with forecasting techniques, particularly as applied to the impact of micro-electronics.

By comparing various methods and models - some of which were national, as in the case of Austria, Canada, the Federal Republic of Germany, Great Britain, Sweden, and others, like that of the ILO, the meeting saw no international solution to the basic problem: What relationship should be established between the information available at plant or sectoral level and the macro-economic data? Until this question is answered, the various studies on the impact of micro-electronics will continue to reach radically different conclusions.

Although it was not the actual subject of the seminar, a majority of the participants believed (albeit on the basis of available methodologies), that the impact of micro-electronics in the medium term may not be as dramatic as some recent studies have suggested (see various articles published in *S.L.B.* since 1979 in the chapter devoted to the effects of new technologies).

The documents and proceedings of the seminar will be published next year.

Source: ILO.